Dragonflies & Damselflies of Cape Cod

Text by Virginia Carpenter

Illustrations by Terry Ellis

The Cape Cod Museum of Natural History
BREWSTER, MA

— NATURAL HISTORY SERIES No.4 —

Published by The Cape Cod Museum of Natural History, Brewster, Massachusetts 02631

The author would like to thank Terry Ellis for her beautiful illustrations which make this book a useful tool for the amateur naturalist, and for her incredible patience during long hours of work.

My thanks also to Robert Finch for editorial expertise and the know-how necessary to guide this project from beginning to end.

I would like to acknowledge Jerrell Daigle for reading and critiquing the manuscript; Sidney Dunkle for comments on various species and for providing the list of English names for dragonflies; Paul Miliotis for help and advice during the early years of my Cape Cod research; Elizabeth Taylor, who accomplished the monumental task of entering my typewritten manuscript into the computer and endured numerous rounds of corrections before the job was done; Bob Power and Susan Lindquist for years of encouragement; and the Museum's Publications Committee for their guidance.

Library of Congress Cataloging-in-Publication Data

Carpenter, Virginia
Dragonflies & damselflies of Cape Cod/text by Virginia Carpenter/
illustrations by Terry Ellis. — 1st ed.
80 pp. — (Natural history series : no. 4)
Includes bibliographical references and index.
ISBN 0-916275-00-2 (pbk.) : $9.95
1. Dragonflies—Massachusetts—Cape Cod—Identification.
2. Damselflies—Massachusetts—Cape Cod—Identification.
3. Odonata—Massachusetts—Cape Cod—Identification.
I. Ellis, Terry (Teresa C.) II. Title: Dragonflies and damselflies of Cape Cod.
III. Series: Natural history series (Brewster, Mass.) : no. 4.
QL520.2.U6C37
595.7'3309744'92—dc20 91-15187

First Edition, May 1991

Printed by Shank Painter Printing Co. Inc., Provincetown, MA

Contents

Today I saw the dragon-fly
Come from the wells where he did lie.

An inner impulse rent the veil
Of his old husk: from head to tail
Came out clear plates of sapphire mail.

He dried his wings: like gauze they grew;
Thro' crofts and pastures wet with dew
A living flash of light he flew.

THE DRAGON-FLY
by Alfred Lord Tennyson

Foreword

As my years of dragonfly-watching progress, I become more captivated by them, more awed by their vibrant colors, exuberant activity and unsurpassed command of the air. I am often asked: why dragonflies? Indeed, this is a difficult question to answer, but I cannot help thinking that I am not alone in the joy I feel when the Long-legged Green Darner graces a pond with his presence. I think of the fisherman who must certainly notice the brilliant blue and black damselflies zipping about the pond's surface; the child whose eye wanders from his bullfrog quarry as a 3½ inch Darner cruises by; and the birder who realizes with a start that the dragonflies on the marsh are nearly as large as the birds she came to watch! When we meet these valuable insects up close, absorb a basic knowledge of their natural history and curious behaviors, we also gain a great respect for them and their prominent place in the complex biological web of freshwater ecosystems.

In the following pages I hope to introduce you to these wonderful insects and to aid in your understanding of who they are and how they fit into the natural diversity of Cape Cod. Although I have described only those species known from Cape Cod, many of these dragonflies are also found throughout southern New England (including Nantucket and Martha's Vineyard) and many of the eastern states. I invite you to take a few minutes on a warm, sparkling summer day at a nearby pond, to witness the incredible variety of dazzling colors and spirited behaviors of the resident dragonflies. I hope you will become as enchanted by them as I am and, with this book in hand, begin to learn their identities.

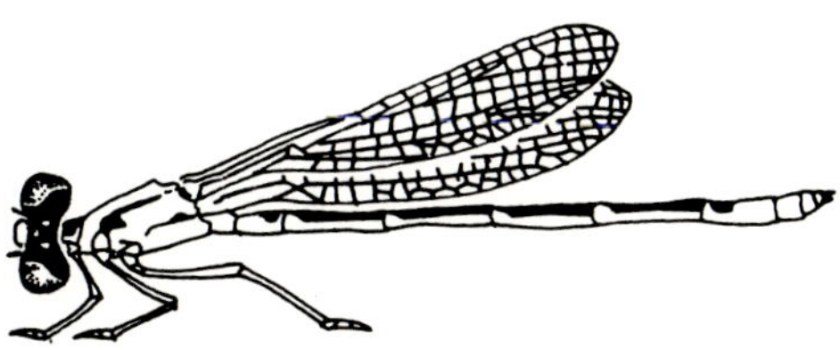

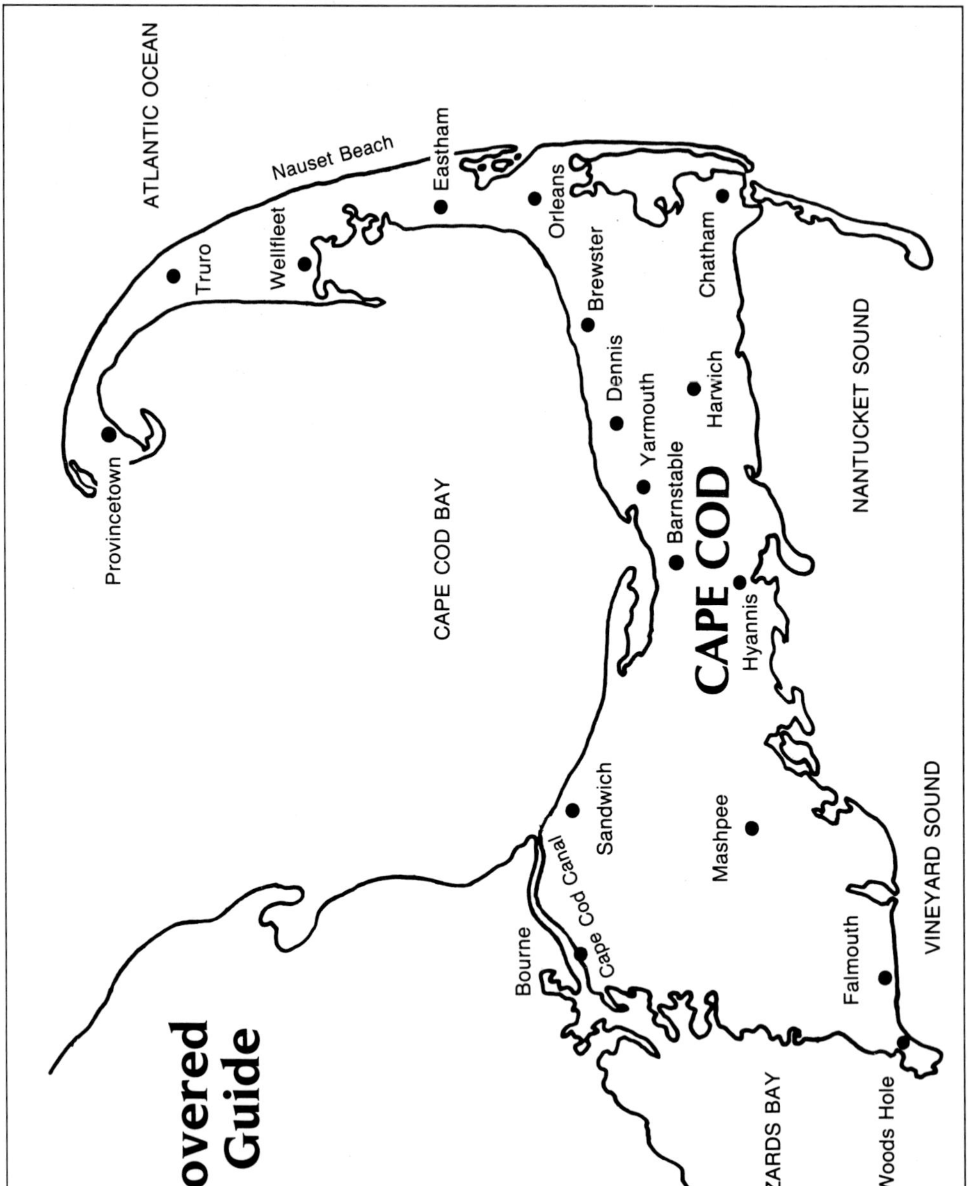

Area Covered by this Guide

Introduction

Dragonflies and damselflies are perhaps the most elegant creatures in the class Insecta. Their large size, resplendent colors and ceaseless activity make them a particularly intriguing group of insects to study. Adults, in general form, are of simple design: exceedingly long, slender abdomen, four elongate netted wings, huge bulging compound eyes and short antennae. Though they spend a good part of their adult life on the wing, members of the order Odonata are inherently tied to aquatic habitats and must return to water to reproduce. Dragonflies and damselflies live a dual life, the first stage as a camouflaged aquatic nymph, the second as an active flying adult.

Basic similarities in form, behavior and evolution have placed the dragonflies and damselflies in the same insect order. But upon close inspection, the two are different creatures indeed. Damselflies, much smaller and more diminutive than the dragonflies, are members of the Sub-Order Zygoptera ("similar wings"). Most damselflies, when at rest, fold their wings neatly against the abdomen or slightly above it (Figure 1). Adult damselflies are more retiring, weaker fliers than their relatives, and most confine their activities to low altitudes well within reach of the shelter of vegetation. Damselfly nymphs are small and slender of body, with three feather-like tails at the tip of the abdomen.

Dragonflies belong to the Sub-Order Anisoptera ("unequal wings"). They are generally larger, more robust insects, powerful aviators which often frequent lofty altitudes. At rest, the dragonflies hold their wings horizontally like those of an airplane (Figure 1). Their eyes are enormous, covering almost the entire head. Nymphs are short, stocky creatures which move forward with a halting motion.

Adult dragonflies and damselflies vary considerably in size, but most are large enough

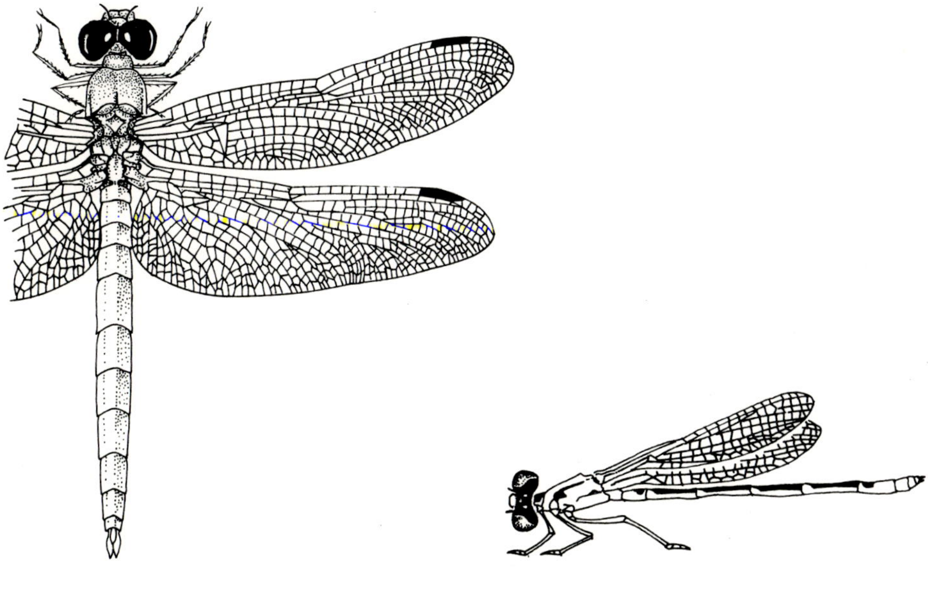

Figure 1. Typical adult dragonfly (left) and damselfly (right).

to be noticed even at a distance. North American dragonflies range in size from the tiny Bluebell (N*annothemis bella*) at only ¾ inch to the behemoth Hero Darner (E*piaeschna heros*) at more than 3½ inches in length.

One of the most delightful and striking features of the Odonata is the variety of color and pattern seen in adults. They are as variegated as any of the birds and butterflies. Almost any combination of color is likely, from rich magentas to exotic sapphire blues to fiery crimson reds. Body colors may be dull and flat or shining and iridescent. They may be painted with leopard-like spots or tiger-like stripes. Even wings are often distinctively marked with dazzling spots or bands of color.

The word Odonata means "toothed", and it refers to the rather formidable serrated mandibles of the adults, one of their main tools for securing food. The common name dragonfly, once applied to both damselflies and dragonflies, has apparently arisen from an age-old fear of these insects. Dragonflies are the legendary "devil's darning needles" and "snake doctors" rumored to be capable of stinging, sewing up ones lips, ears and hair, and making dead snakes come alive. None of these tales are true, of course, but are merely stories passed from generation to generation. The name "dragonfly" and its associated superstitions may also have arisen from the truly fierce expressions of these insects at close range. Beauty, however, is in the eye of the beholder.

As an insect order, the Odonata is a relatively small group. Coleoptera, the beetles, contains over 250,000 species worldwide. Lepidoptera, the butterflies and moths, contains 125,000 species worldwide. Odonata, the dragonflies and damselflies, encompasses only about 5,000 species worldwide, 450 of these in North America, and approximately 180 in Massachusetts. On an even more local level, about 95 species are known from Cape Cod. Their numbers and diversity are greatest in the tropics where year-round warmth and dense prey resources promote a broad and varied fauna. Yet even in far northern tundra regions there are representatives of this insect group, though their diversity there is low.

The Past

Dragonflies and damselflies represent some of the Earth's most primitive, yet incredibly successful insects. Long before the dinosaurs prowled the Earth, ancient winged giants fully 2½ feet from wingtip to wingtip sailed over Paleozoic coal swamps. Dying, these gargantuan dragonflies left remarkable fossils, netted wings clearly preserved in 250 million year old limestone. Through time, world-wide tropical climates cooled and the birth of harsh seasons brought changes in the Earth's flora and fauna. Today, even the largest dragonflies are but scaled-down models of their ancestors. North America's largest dragonfly has a wingspan of 4¾ inches, while some South American species are as large as 7½ inches from wingtip to wingtip.

Today's dragonflies remain primitive in form, having retained several features of their ancient ancestors. Their inability to fold their wings neatly over the abdomen is one such archaic characteristic, as is the dense network of veins which give strength to these wings. More evolutionarily advanced insects, such as butterflies, have far fewer veins in their wings, and exhibit a much different method of flight. Unlike some modern insects which may beat their wings an astounding 1,000 times per second, dragonflies are capable of only 30 beats per second. Furthermore, when a butterfly takes to the air, all four wings move in unison, up and down together. But when a dragonfly launches itself from a perch, its front wings rise as its hind wings drop. In fact, those front wings, as they move, generate tiny whorls of turbulent air about their tips. These tip vortices are then put to use by the hind wings, which transform potentially dangerous unstable air into lift producing forces. Paradoxically, all of these comparatively "ancient" flight mechanisms combine to give unsurpassed stability, speed and mobility to dragonfly flight. In fact, the "unsteady aerodynamics" of dragonfly flight have intrigued modern scientists, who are studying the form and function of these insects, hoping to understand their flight secrets and apply them to the design of our own flying machines.

Life History

Most of us are familiar with dragonflies and damselflies as colorful winged adults. We have seen them performing their acrobatics over ponds or resting on a convenient perch. As adults they present themselves to us, offering us close-up glimpses of their sparkling colors. But the adult stage is only part of their story. To view the beginning stages of their lives we must go in search of them, for the larvae are aquatic and secretive, living hidden beneath the surface of freshwater ponds, lakes, rivers or bogs. Unlike the adults they do not present themselves to us in a myriad of brilliant colors and riotous activity, but lead much different lives as part of the intricate biological web beneath the water's surface.

One of the most interesting aspects of insect life is the miraculous change, or metamorphosis, which they endure in the process of "growing up" to adulthood. More modern insects such as butterflies, bees and beetles, pass through four stages (egg, larva, pupa, adult), each drastically different from the previous stage. But dragonflies, true to their ancient origins, pass through only three stages before attaining the ultimate event of all life, reproduction. This simple, three-part metamorphosis is typical of more primitive insects including bugs, mantids, cockroaches and dragonflies. After emerging from the egg, the insect progresses through a series of molts as a larva before finally reaching the adult stage.

Oviposition and Eggs

Life begins for all dragonflies and damselflies as tiny eggs deposited in freshwater habitats. The eggs vary in shape and size depending on the method and place in which they are deposited. Most are minute, measuring less than one millimeter in length.

Dragonfly eggs are laid in one of two ways. They may be dropped freely into the water, as is true of most dragonfly species, or they may be carefully placed inside the leaves or stems of aquatic plants. This latter method is typical of damselflies and some dragonflies which are equipped with sharp knife-like ovipositors for slicing into fibrous plant tissues.

Some of the most intriguing adaptations in egg morphology are seen in those species which breed in flowing water. Obviously, the eggs must be secured against the force of the stream's current or they will be swept away and lost. And so, in river dwellers such as the Clubtails (Family Gomphidae), the eggs possess ingenious mechanisms to ensure survival. Some bear rather curious coiled filaments which unravel instantly on contact with water to entangle with aquatic plants. Others are encapsulated in a gelatinous adhesive which serves as a water-proof "super-glue" in securing eggs to stable objects.

The process of egg-laying in insects is called ovipositing. It is a fascinating behavior, easily observed at almost any Cape Cod pond; and, as with so many aspects of dragonfly life, it varies dramatically from species to species. In most cases the female is guided to a suitable spot by a male, who may remain physically attached to her or hover nearby like a sentinel guarding a prized jewel. This behavior, called mate guarding, arises from the fact that in dragonflies the last male to mate with a female will fertilize her eggs. If he allows another male to copulate with her, his genetic future with this female is lost. Thus, territoriality and mate-guarding have evolved in many dragonfly species to ensure that the guarding male will be the last to mate with the female. If you have spent time on a summer day at a local pond, you may have observed paired dragonflies skimming along the surface of the water or perched in the shelter of aquatic plants, the male in front, the female behind. This paired flight is called tandem flight, and it is always associated with breeding, especially in the Zygoptera (damselflies). After mating, many of the Anisoptera (dragonflies) do not remain attached, and the female goes about the act of egg-laying alone.

If the male remains attached to the female throughout oviposition, he will guide her to the appropriate spot for egg-laying. In damselflies which deposit their eggs inside plant tissues, the female will back down into the water, clinging to the stem of the plant for support (Figure 2). At times she may be totally submerged during oviposition, and may remain below the water's surface for as long as 30 minutes. In contrast, most dragonflies deposit eggs unprotected, directly into the water. Females may hover several feet above the surface dropping eggs from the tip of the abdomen, never actually touching the water. Or they may fly low, within a few inches of the pond, tapping the surface repeatedly with the abdomen in a gentle rhythmic motion, releasing eggs with each contact (Figure 3). These eggs then sink to the bottom sediment or adhere to submerged plants. In stream-dwelling Clubtail dragonflies, eggs are often deposited directly on or in the bottom sediment. The female plunges her abdomen into the sand and releases eggs in this manner.

Tandem flight during oviposition is not as frequently observed in dragonflies as in damselflies. However, particularly in Common Skimmers (Family Libellulidae), a male is often nearby guarding the ovipositing female, making bold dashes at intruding males. Apparently if competition from other

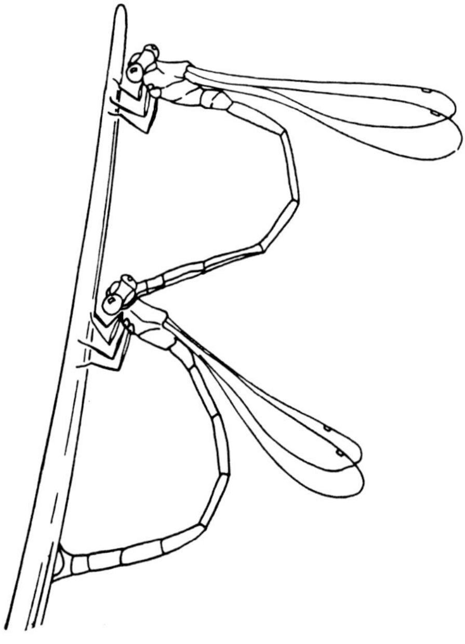

Figure 2. Damselfly pair ovipositing, female below.

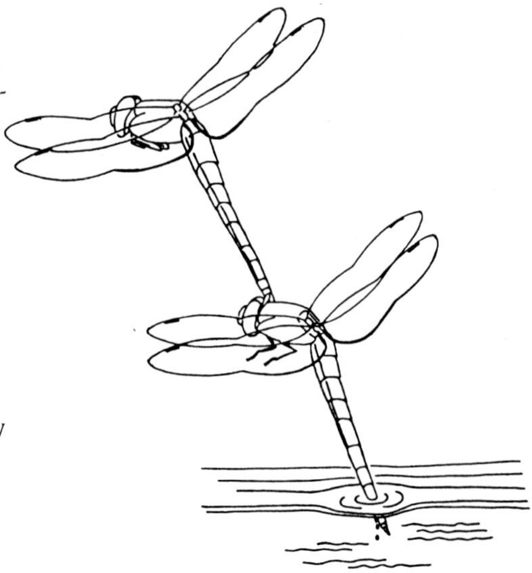

Figure 3. Dragonfly pair ovipositing, female below.

males is intense, the guarding male will be more attentive to his mate. Male Whitetails, (*Libellula lydia*), are especially well-known for their mate-guarding behavior, but will also wander from one female while in the act of guarding her to mate with and guard another!

Insects are incredibly productive animals, compensating for heavy mortality by laying tremendous numbers of eggs. The reproductive potential of dragonflies varies with species, but the number of eggs deposited in individual batches ranges from about one hundred to well over 5,000. Some females, if they live long enough, may oviposit several times. Eggs begin to hatch within a few weeks of oviposition in some species, but in others they may overwinter before hatching.

The Larva

The dragonfly larva is called a nymph. It is aquatic, spending varying amounts of time beneath the surface of fresh or brackish water. Like other insects, dragonfly nymphs progress through a series of molts from the tiny prolarva which emerges from the egg, to

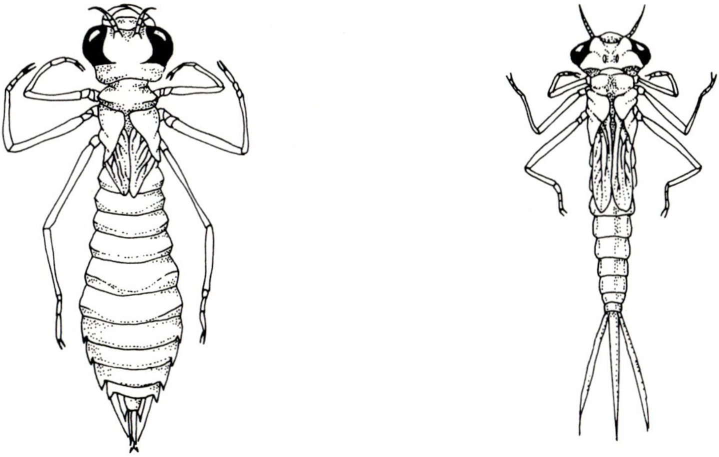

Figure 4. Dragonfly nymph (left) and damselfly nymph (right).

the fully grown nymph ready to leave the water. Insects are encased in a rigid protective shell called an exoskeleton, which does not stretch with the growing larva. Thus, they must shed this covering in stages so that growth and development may continue. In dragonflies the number of molts ranges from 8 to 16.

In both appearance and behavior, the dragonfly nymph is a very different creature from the adult. It is a somewhat retiring, drab insect hardly resembling the active colorful adult it will become. Yet both adult and nymph are ravenous predators consuming enormous quantities of living prey, often as large as themselves. Physically the dragonfly nymph is a short, rather dumpy creature with long sprawling legs (Figure 4). The damselfly nymph is more slender and graceful in appearance, its most distinctive feature being the three featherlike "tails" at the end of the abdomen (Figure 4). Both dragonfly and damselfly nymphs are equipped with a peculiar apparatus for securing food. The

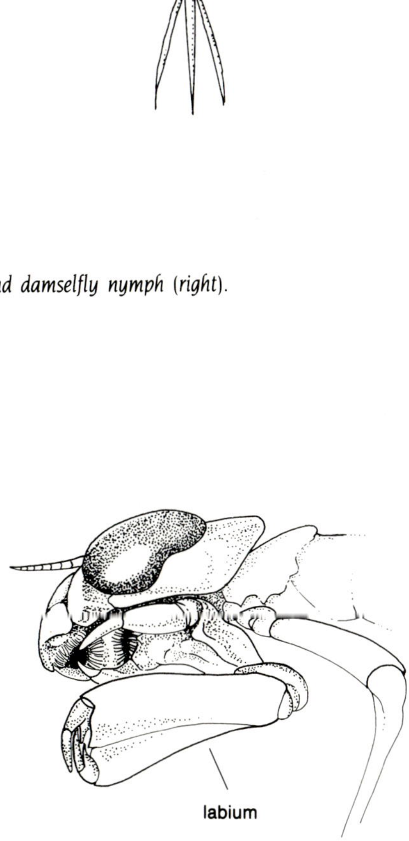

Figure 5. Head of dragonfly nymph showing labium.

lower lip or labium, is a moveable, hinged mechanism which is normally folded beneath the head of the nymph (see Figure 5), hiding the mouthparts like a mask. The neat way in which the labium is stored belies its true function in capturing prey. When a food item is within reach the labium is projected forward at lightning speed, and food is grasped tightly with two terminal hooks. This physical structure is unique to the Odonata.

Dragonfly larvae breathe by means of internal rectal gills. Damselflies on the other hand, possess a secondary mechanism for oxygen absorption in the form of three feathery tail-like appendages. These caudal lamellae also give the damselfly its elongate, slender appearance and are a major distinguishing feature in separating damselfly and dragonfly larvae.

As the larva grows, molts and matures, evidence of its future form becomes more noticeable. On the nymph's back four flattened wing-pads develop, which will ultimately provide the adult with its most important capacity, flight.

In addition to the significant physical differences between dragonfly and damselfly nymphs, there are also more subtle differences among the species. In fact the nymphs of various species are as distinct as the adults. Body shape, size, leg length, and behavior vary from species to species, many of these characteristics being special adaptations to a particular habitat or way of life. The most conspicuous examples of this are the larvae of species which live in flowing waters and are compressed dorso-ventrally to allow hiding in crevices and to facilitate movement in currents. The burrowers, which tunnel through bottom sediments, are typically short-legged, long-bodied creatures.

Feeding and Behavior

Dragonfly nymphs generally lie in wait for their prey, hiding motionless in aquatic vegetation or bottom muck. Food is located through both visual and tactile means. The nymphs may sense movement of prey with their antennae or legs. The legs of some nymphs are unusually long and spidery, which may facilitate a greater range of prey perception. Regardless of how food is located, the technique employed for capturing it is ambush. Generalized obligate carnivores, Odonate nymphs consume virtually anything that ventures near enough to be caught. They have been observed feeding on a variety of aquatic insects, crustaceans, worms, even small fish and tadpoles. They are also known to be cannibalistic if food resources are low. On the other hand, the nymph is itself an important food source for aquatic animals including spiders, other insects, fish, frogs and turtles. Of these predators, fish are most significant, and on Cape Cod the voracious pickerel (pike) and bass are major consumers of Odonate nymphs. In fact, when fish populations are artificially increased for sport purposes, Odonates may suffer.

While much larval time is spent resting immobile, Odonate nymphs are quite capable of moving about, both at the water's surface and on the bottom. Dragonfly nymphs often walk over the bottom in search of prey, their legs being the primary mode of locomotion. Or they may move by a more unusual method: jet propulsion. By intaking and expelling water through the rectum, Anisoptera larvae effectively jet through the water. The dragonfly nymph moves rapidly, though somewhat jerkily, as water is ejected from its body, and further enhances its less-than aerodynamic form by pressing its legs tightly against its abdomen. This proves to be an excellent escape mechanism when the nymph is threatened.

Damselfly larvae exhibit a different mode of transportation, one which is considered more primitive than that of the dragonflies. They too frequently walk; but when danger threatens damselflies also swim: a lovely sinuous side-to-side motion not unlike the movement seen in snakes and salamanders.

Both dragonfly and damselfly nymphs spend a great deal of time resting as a method of concealment. There are generally two categories of resting behavior depending, upon where the nymph chooses to hide and the position it takes once there. Not surprisingly, the shape of the nymph has become adapted to the particular behavior it employs.

Some larvae are burrowers, and these species either dig into bottom sediments or hide beneath muck and debris. On Cape Cod, the Clubtail Sand Dragon (*Progomphus obscurus*), is a good example of a burrower. Nymphs of this species plow along through the sand like miniature moles, leaving meandering trails as evidence of their presence. These trails are easily spotted along the shallow shorelines of most clear sandy Cape Cod lakes. When disturbed or uncovered these short-legged nymphs dig frantically, covering themselves with a shallow blanket of sand. One can usually assume by the short stocky legs of certain species that they are burrowers. Many Common Skimmers (Family Libellulidae), which inhabit muddy ponds heavily littered with decaying material, simply cover themselves with debris and are thus effectively concealed. These nymphs are often revealed when a net is carefully drawn along the bottom of murky ponds. At first no movement may be detected in the muddy contents of the net. But as one pokes through the ooze, subtle movements can be seen: that of a leg here, an abdomen there. Burrowers such as these remain motionless at first and if picked up will tuck their legs tightly against their bodies in an effort to look as little like a living thing as possible. Curiously, in some bottom-dwelling hiders such as the River Cruiser (*Macromia illinoiensis*), the legs are extraordinarily long and the abdomen rather round and short. These long legs probably act as sensory receptors, aiding the hidden insect in perceiving the movement of potential prey.

The second method of concealment is typical of most damselflies and some dragonflies (darners). These nymphs simply cling to a submerged object for camouflage. They do not attempt to cover themselves with bottom trash. The dense tangles of rushes, grasses and other aquatic plants in the more shallow Cape Cod ponds provide excellent underwater camouflage for these nymphs. Many species which hide in submerged jungles are capable of changing color to enhance their camouflage. Others, such as Green Darner larvae, are cryptically patterned with bands of pale green and brown which render them virtually invisible in the weedy places of the ponds.

Emergence

Odonate nymphs mature in varying amounts of time depending on species, food supply, temperature and habitat. The Globetrotter (*Pantala flavescens*) requires less than two months to mature and even the Green Darner (*Anax junius*) is capable of maturing in just four months. Some slow growing species, on the other hand, may require several years to develop. Inhabitants of vernal (temporary) ponds must grow rapidly and be ready to take to the air before the habitat dries up. On Cape Cod, it is likely that most dragonflies develop fully from egg to adult in one to three years, thus passing at least one winter in the aquatic stage.

Emergence in dragonflies and damselflies is a dangerous time. The task itself is an all consuming and arduous one, resulting in a beautiful and astonishing transformation that is fascinating to observe. The fully mature nymph seeks a support of some sort, usually emergent plants, rocks or man-made structures protruding above the water's surface. I have observed exuviae (cast off nymph "skins") clinging to docks, boats, rocks and even floating trash. In the absence of a support nymphs creep away from the water to the nearest suitable spot. Obviously, the shorter the distance the insect travels the less energy it expends and the less vulnerable it is to predators, which can cause heavy mortality during emergence. Nymphs have been known to travel over 100 feet to an appropriate support, but most stay within a few feet of the water.

Some species, including the big darners (Aeshnidae), emerge before first light, when the dangers of predation are reduced. Others emerge during the morning hours. Though the nymph has been preparing itself for some time for this final most critical molt, the actual transformation or ecdysis, takes no more than just a few hours.

Once secured to a support, the larval skin splits open on the middle body section or thorax, between the two pairs of wing pads. Then, as though extracting itself from a tight glove, the adult pulls its head, thorax and abdomen from the larval skin. This new or teneral adult is soft-bodied, wrinkled and dull-colored. Its wings, previously crumpled into four larval wing sheaths, are now free to expand as the blood pumps into them. The abdomen, soft and shriveled, must also expand to its full length and harden into a protective suit of armor. It will take the new adult anywhere from a few minutes to several hours to complete emergence, but it will be a few days before its body is fully strengthened and its colors fully developed. For the first hazardous hours of adult life, the dragonfly clings tenuously to its cast-off skin. Its wings, destined to be strong and sturdy for flight, are as soft and fragile as wet tissue, and thus the insect cannot fly. It is at the mercy of the weather and a gauntlet of hungry predators. A gust of wind or a cold rain could easily be disastrous to the dragonfly. Spiders take large numbers of emerging dragonflies and teneral adults. I have often observed empty larval skins and weak flying tenerals caught in the gossamer threads of their webs. Birds too, take advantage of the vulnerable state of emerging dragonflies, snapping up nymphs as they crawl away from the ponds and soft-bodied tenerals on their maiden flights. And in an ironic twist of fate, other dragonflies, still in the voracious aquatic stage, are known to crawl out of the water to eat those in the process of emergence!

Many spring species emerge from ponds en masse, and it is often possible to see evidence of mass emergence in the form of empty skins still clinging to aquatic plants. Close inspection of emergent vegetation usually reveals several of the exuviae, but after a mass emergence it is not uncommon to find hundreds scattered about the circumference of the pond.

When the teneral adult is ready it will embark on its first, or maiden flight, leaving behind forever the watery confines of the pond. Maiden flight is fluttery, weak and tentative, and the dragonfly is conspicuous with its glistening new wings and pale body colors. If you walk along the edge of a pond after a mass emergence you will flush a cloud of teneral adults, reluctant to desert their nymphal skins, yet anxiously diving for shelter in the nearest shrubbery.

Before it is able to reproduce, the teneral adult spends a few weeks away from the breeding habitat, feeding, strengthening and maturing. The larger, stronger flying dragonflies, particularly the darners, may travel as far as 600 feet from the emergence site on the maiden flight and then wander further during the first few weeks of adult life. Most damselflies can be found several hundred yards from the nearest water, having drifted with the wind like aerial plankton on feeble new wings.

In order to fully comprehend the magnitude of the change that has taken place through emergence, we must consider the individual transformations involved. First the aquatic nymph, unobtrusive, dumpy, and sluggish, has become in a matter of hours an active, brilliantly colored winged predator. Furthermore, the nymph, breathing with gills, has evolved into an air-breathing adult with the usual insect system of tubes (trachea) and spiracles for aerating its body. The larval eyes, only part of a well-developed sensory system, have transformed into the enormous compound eyes of the adult, which will depend solely on sight for locating its prey. From small flattened pads on the back of the nymph have emerged four large shimmering wings which will carry the dragonfly through the last and most significant phase of its life. Unlike the nymph which goes to great lengths to conceal itself in the murky underwater environment, the adult dragonfly is a dynamic, agile creature of the sun.

> *Thus the dragon-fly enters*
> *upon a more nobler life than*
> *that it has hitherto led in the*
> *water, for in the latter it was*
> *obliged to live in misery, creeping*
> *and swimming slowly, but now it*
> *wings the air.*
>
> Jan Swammerdam (1637-80)

The Adult

With the task of emergence complete, the adult dragonfly is free to move away from the breeding site to feed and mature. Adult life is relatively brief, lasting only a few hectic weeks in some species to several months in others. In more northern climates, adults will die before the onset of winter.

Dragonflies and damselflies are devoted sun-worshippers, and their activities are vigorous on bright summer days. Before returning to the breeding site adults spend a few days to a few weeks sunning, hawking mosquitos and preparing for the rigors of the reproductive season.

During this time of maturation they will lose their pale teneral hues and develop the full kaleidoscope of adult colors. They will spend their time feeding and loafing over open fields and lawns, along woodland paths and over marshes.

Like other insects, the dragonfly body consists of three major segments: the head, thorax and abdomen (Figure 6). Their bodies are long and lithe, sturdy but graceful in silhouette, the evolutionary culmination of a superb aerodynamic design.

The most notable features of the head are the huge multi-faceted compound eyes which wrap about the head, giving the insect nearly 360 degrees of vision. Adult dragonflies rely entirely on sight in hunting. The enormity of the eyes and relative reduction in antennal length illustrates the importance of vision in these insects. Compound eyes contain several individual lenses, or ommatidia. The number of lenses in insect eyes ranges from 6-8 in some worker ants to an astounding 28,000 in the darner dragonflies. The more

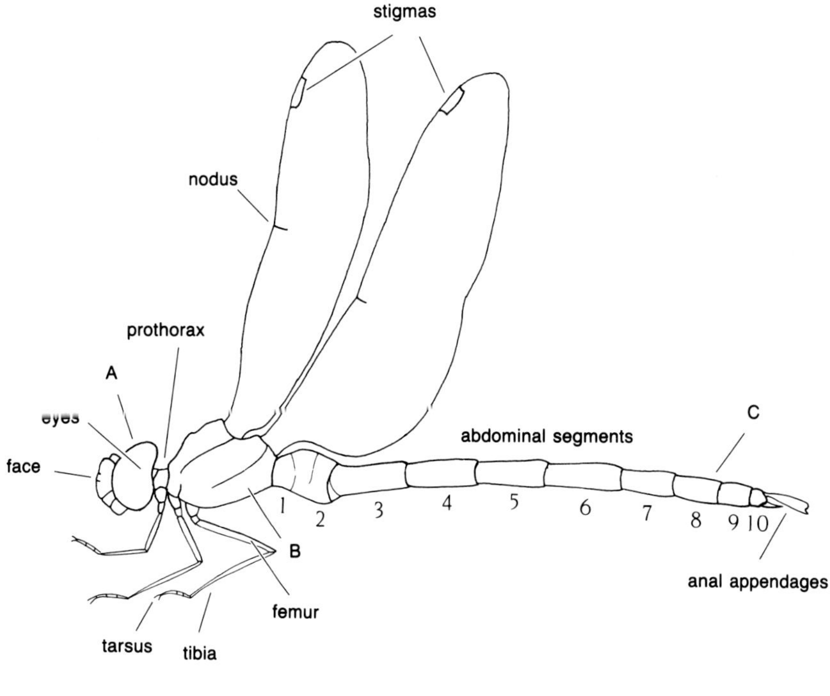

Figure 6. The three major body segments (A: *head;* B: *thorax;* C: *adbomen*) *and body parts of an adult dragonfly.*

lenses in the compound eyes, the greater is the insect's visual acuity. Thus in the dragonflies, whose eyes contain the maximum ommatidial number, vision is exceptional. I have often marveled at their uncanny ability to avoid my net, an ability that is a combination of aerial wizardry and extraordinary vision. In fact, dragonflies are capable of seeing moving prey, predators and collectors at distances of 60 feet and more!

The eyes of dragonflies may be exquisitely colored with greens, blues or reds, which sparkle with the reflection of light on the

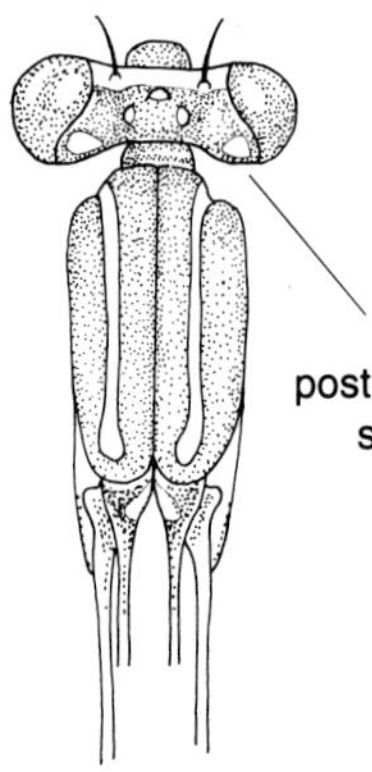

Figure 7. Closeup of damselfly head and thorax showing post-ocular spots.

minute facets of the living insect's eyes, but which fade disappointingly in death. Insects also have three simple eyes called ocelli, located between the antennae. In many of the Zygoptera, there are two distinct spots of pigment, called post-ocular spots, behind the eyes (Figure 7). These are often brightly colored in blue or green, and may be useful in species identification.

Another significant feature of the dragonfly head is its mouth. The actual jaws, which are sharply serrated for efficient maceration of prey, are hidden behind an upper and lower lip, the labrum and labium. While they are quite capable of inflicting a painful bite, dragonflies are generally peaceable creatures when handled carefully.

The most crucial function of the dragonfly thorax is mobility, for it is here that the wings and legs arise. The thorax is divided into two sections, the prothorax, which bears the head and the first pair of legs, and the synthorax, which bears the wings and hind two pairs of legs. Though the thorax is actually further divided, these divisions are more complex and beyond the scope of this book. For the scientist studying dragonflies (called an Odonatologist), knowledge of thoracic morphology is imperative, as various parts contain diagnostic markings valuable in determining species identity.

Perhaps the most important feature of Odonate anatomy are the four membranous wings which carry the dragonfly smoothly and effortlessly through adult life. In the Anisoptera the hind wings are larger in area than the forewings, while those of the Zygoptera are of approximately equal size and shape.

Dragonfly wings are strengthened and braced by a labyrinthine network of veins and cells. It is this venation, in part, which gives the wings their shimmering jewel-like appearance under the sun's reflected rays. Although it is not necessary for the casual observer to be familiar with all wing veins, a basic knowledge of venation is both interesting and useful in identification of various family groups (Figure 8). Odonatologists of course, have an in-depth understanding of wing venation, since it is an important tool in species identification. A working knowledge of Odonate wing veins would include familiarity with the anal loop of the hind wing, the triangles, stigmas and nodus. Figure 9 shows the major veins and cells of the Odonate wing, many of which are easily seen with the insect in hand.

Most of us who have seen dragonflies at close range have noticed the four colored cells, one on the leading edge near the tip of each wing. These are the stigmas, which serve as blood pools giving vital stability to the insect in flight. The stigmas may be black, red, orange or white, or they may be bi-colored; but they are almost always distinctly different in color from the rest of the wing.

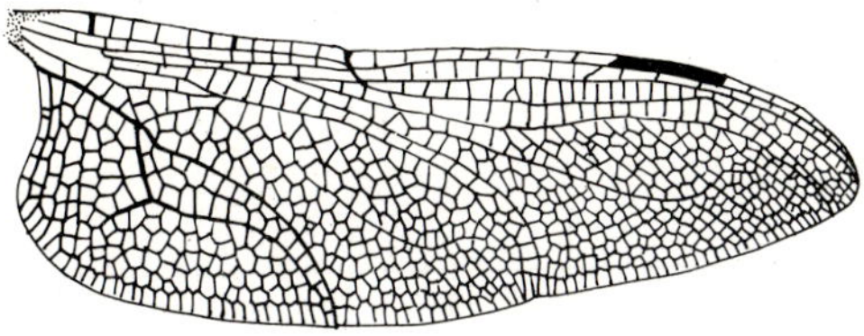

Family Aeshnidae
(Darners)

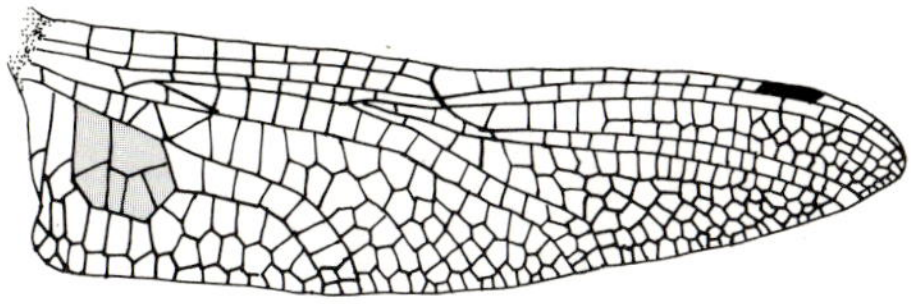

Family Macromidae
(Cruisers)

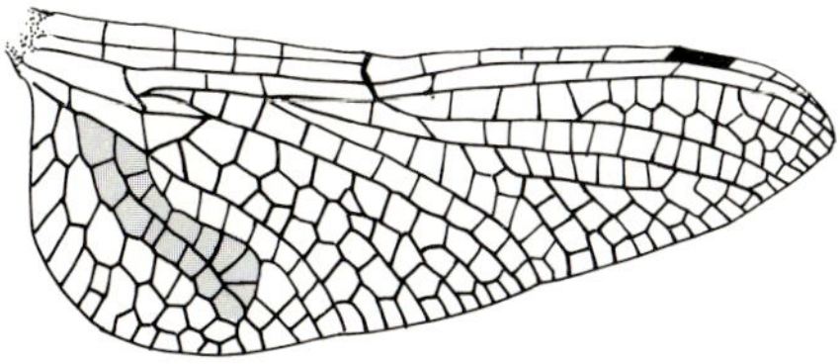

Family Corduliidae
(Emeralds)

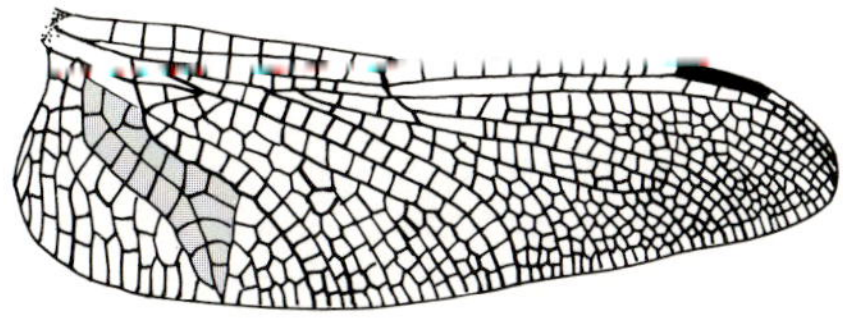

Family Libellulidae
(Skimmers)

Figure 8. Dragonfly hindwings of four dragonfly families showing characteristic venation patterns. Anal loops are shown shaded.

Figure 9: Major veins and cells of the wings of a Gomphidae (Clubtail) dragonfly (left) and damselfly (right). Anal loop in dragonfly is shown shaded.

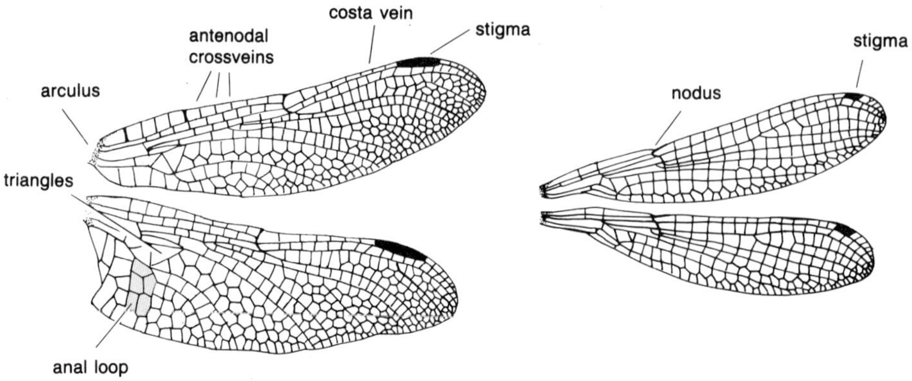

The wings of most dragonflies are transparent or hyaline, but a broad spectrum of color and pattern exists. The Spot-winged Skimmers or Pennants (genus *Celithemis*) all have handsomely patterned wings, some embellished with red spotting, others with orange or black spotting. Perhaps the most ornate color is seen in the Tenspot (*Libellula pulchella*), which boasts a harlequin black and white checkerboard pattern on each wing. Other species are adorned with a single color throughout the wing membrane, such as the Goldenwings (*Libellula auripennis*) and the Amberwings (*Perithemis tenera*) with tawny orange wings, and the Ebony Jewelwing (*Calopteryx maculata*) with velvety black wings.

The dragonfly's legs, not particularly useful in walking, are instead most efficient for perching and snaring prey. In fact the legs are positioned on the thorax so they face forward, an arrangement which facilitates both perching and the capture of prey. The legs, heavily armed with spines, are held like a basket in flight, a formidable and deadly trap for other insects. Dragonfly legs consist of several segments which are illustrated in Figure 6.

The third body segment, the abdomen, is exceedingly long in dragonflies, resembling a tail or, to some, a stinger. The Anisoptera abdomen is generally stockier than that of the Zygoptera, which is needle-thin and delicate in appearance. The extremely slender Zygoptera abdomen has earned the damselflies the name "darning needle".

The Odonate abdomen comprises ten segments, each with a pair of circular openings called spiracles through which the insect breathes. The abdomen also bears, in the female, the egg-laying apparatus and, in the male, two sets of appendages necessary for mating and reproduction (Figure 10). The presence of these features is the most certain determination of sex in adult dragonflies.

The female reproductive apparatus is located near the tip of the abdomen on the ventral surface of segments 8 and 9 (Figure 11). The ovipositor on segment 8 (present in Zygoptera and some Anisoptera) and the vulvar plate are variable in shape and size, and are thus important in species identification. Dragonflies which lay eggs inside plant tissues have well-developed ovipositors, whereas those which drop their eggs freely into the water may lack the structure completely. The vulvar plate attains its most peculiar form in female Emeralds (*Somatochlora* sp.) where it is often quite long, projecting downward and perpendicular to the

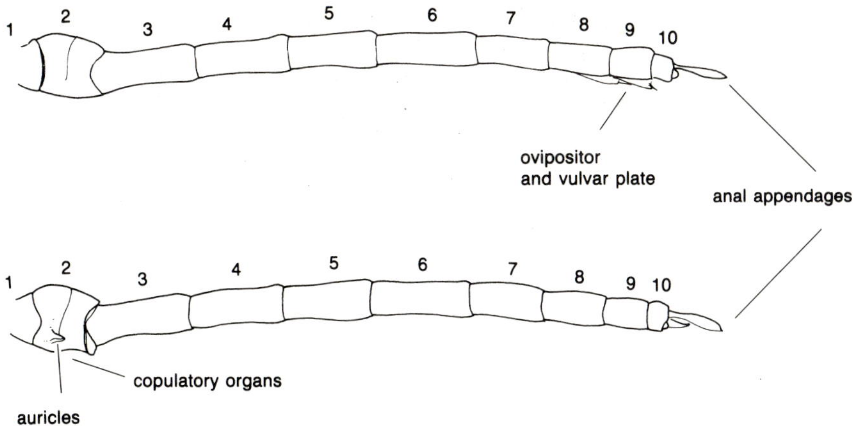

Figure 10. Abdominal segments and appendages of female (above) and male (below) dragonflies.

abdomen. This scoop-like plate is designed for laying eggs in wet sand or mud. The tenth segment of the female abdomen bears a pair of anal appendages, which may be inconspicuous in damselflies but are quite well-developed in dragonflies.

Male appendages are a bit more complex, but like those of the female they are distinctly different from species to species, and therefore excellent identification characteristics. They consist of claspers on segment 10 and accessory or copulatory organs on segment 2 (Figure 10). Curiously, sperm is produced in an organ near the tip of the abdomen and must be transferred prior to mating, to the copulatory organs on segment 2. To accomplish this, the male simply bends his abdomen down and forward in a sort of circle. A spermatophore can then be transferred to the female via the accessory organs on segment 2. Anal appendages on segment 10 consist of a superior (upper) pair and single or paired inferiors (lower) which, like claspers, serve to hold the back of the female's head or thorax during mating (Figure 11). The male appendages are designed to fit precisely into grooves behind the female's head. The coupling of these appendages is a critical lock-and-key mechanism, insuring that only males and females of the same species mate. The system does on occasion break down and attempted mixed species pairings do occur. In one particularly bizarre case, I observed two male Bluets (*Enallagma* sp.) of different species attempting to mate!

The color pattern of the Odonate abdomen is as spectral and varied as that of the rest of the body. Generally, as in most birds, males and females are colored differently, the male being considerably brighter than the female. As the dragonfly ages its colors change also. Many male Common Skimmers (Family Libellulidae) acquire a powdery bluish-white abdominal coat called pruinosity. This flashy white color is used in territorial displays between competing males and as a recognition clue between the sexes during courtship. Indeed, abdominal colors in many Odonates play a significant role in recognition between individuals of the same species and those of different species.

Flight

Dragonflies and damselflies are superbly designed winged predators, much like the peregrine falcons of the class Aves: slender and streamlined in form, agile and maneuverable on the wing. Despite a relatively primitive anatomy and decidedly ancient flight mechanism, the design of the dragonfly has been tried and tested through

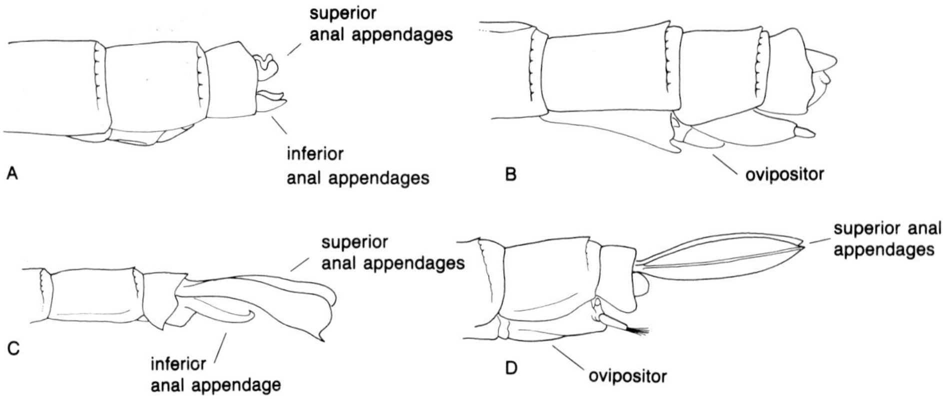

Figure 11. *Detail of abdominal appendages in male and female* Odonates: A. *male damselfly* (Enallagma recurvatum); B. *female damselfly* (E. recurvatum); C. *male darner* (Aeshna umbrosa); D. *female darner* (A. umbrosa).

250 million years and has emerged a most successful flying machine.

The unusual dynamics of dragonfly flight enable them to perform with unsurpassed prowess in the air. Larger species such as the darners are capable of attaining speeds of 30 miles per hour and more, their strength and balance perfected to the point where prey fifteen times their own weight can be lifted and carried off. One of the most critical aspects of dragonfly flight however, is the remarkable maneuverability which allows them to snatch moving prey out of the air. Dragonflies are one of the few insects which are able to fly in almost any direction: forward, backward, sideways, up and down, and even upside down! They stop with incredible abruptness to hover like miniature helicopters, almost completely stable in the air except for a barely perceptible rocking motion of their bodies. While the Anisoptera are powerful and swift on the wing, the more diminutive Zygoptera are more tentative and sluggish.

Dragonflies depend upon their wings for nearly all aspects of adult life: feeding, aggression, reproduction and migration. Even mating takes place in flight. Adults must be able to capture enough food to sustain themselves, avoid being gobbled up by hungry birds, defend a territory, support and carry a female during mating, and move to new habitats when the old becomes inhospitable . . . all of which is accomplished in flight.

Behavior

Dragonflies spend much of their adult life feeding, patrolling and defending a territory. Most species are active by day, but many feed at dusk and a few are nocturnal. Adult activity is furious on the sunny warm days of summer, but when clouds block the sun adults seek shelter. Because they are cold-blooded, dragonflies must raise their body temperature by sunning before launching into a day's activities. If the day remains cloudy, windy and cold, the pond shore, usually teeming with activity, is strangely silent. Very high temperatures and humidity also quiet the activity of dragonflies. In these conditions, adults cool themselves by tilting their wings slightly to shade the thorax or by adopting a near headstand posture to expose as little surface area as possible to the baking sun.

Adult dragonflies are generalized obligate carnivores with seemingly insatiable appetites, taking any food item that is small enough to be carried and slow enough to be captured. Prey is caught on the wing, approached and seized from below and in most instances consumed in flight. Unusually large insects are carried to a convenient perch and eaten at rest. The bulk of the Odonate diet consists of flies and other Dipterans as well as unwary damselflies and dragonflies. Many preferred food items are those considered undesirable pests to humans, including green-head flies, deer flies, horseflies, mosquitos and biting midges ("no-see-ums"). The dragonflies penchant for mosquitos has earned it the colloquial name "mosquito-hawk".

Just as the dragonflies are voracious predators, they too fall victim to other equally ravenous animals. The most successful predators of adult dragonflies are birds, from herons to flycatchers. Weak-flying damselflies fall prey to birds more frequently than do dragonflies, but both are susceptible. Fish and frogs also take adults, chiefly those flying low over the water or ovipositing, and spiders snare teneral adults in their pond-shore webs. Perhaps the most unusual predator of dragonflies on Cape Cod is the carnivorous Sundew plant (*Drosera* sp.), which is common around our Coastal Plain Ponds. Damselflies in particular are trapped in the sticky sweet fluids of Sundew leaves and digested by the plant as it supplements its nitrogen supply. Odonates, both nymph and adult, are thus a principal link in the intricate, interconnected food web in aquatic and terrestrial ecosystems.

Like most animals, dragonflies are also the victim of parasites. The most noticeable Odonate parasite is a tiny red water mite (*Hydracarina*) which is often found in clusters adhering to the thorax or abdomen. The intertwined lives of these two animals are remarkable: the larval water mite attaches itself to the aquatic Odonate nymph, sucking on the host's body fluids for nourishment. When the Odonate emerges the mite moves onto the adult. During oviposition it drops back into the water to complete its life.

Feeding behavior in adult dragonflies is a wild and frenzied activity, almost uncontrolled in appearance, particularly when the dragonfly is working a large swarm of prey insects. Quick, erratic directional changes and rising spiral flight within the swarm allow efficient pursuit and capture of mosquitos and midges. This behavior is frequently seen at twilight on warm humid nights when dense clouds of mosquitos and midges paint the air.

At times large numbers of dragonflies congregate where prey are concentrated. It is not uncommon to encounter dozens of darners hawking (i.e. chasing) mosquitos in fields, backyards, or along sunny roads. I once spent an hour in a Provincetown field attempting to net Globetrotters (*Pantala* sp.) as they coursed steadily and regularly across the field, making quick upward stabs at their prey. Their agility and speed were mesmerizing and all of the 20 or 30 Globetrotters managed to avoid being captured! At times, however, not all of their predatory behavior seemed in earnest. On numerous occasions the Globetrotters would snare a fly, the contact of predator and prey producing an audible "click", and then drop it as though it were distasteful!

Many male dragonflies are territorial, a behavior which is primarily associated with mating and reproduction. Most territories are transient, the dragonfly occupying them only for a brief period of time. Territorial defense is vigorous when large numbers of males are on the habitat. The area is defended pugnaciously from other males but left open for the approach of a female. Within the territory a male may have one or more favorite perches from which he surveys his kingdom. I have often watched individuals return repeatedly to the same branch or rock after a brief chase or hunting mission. Some of the larger dragonflies, the darners in particular, rarely perch below treetop level.

Males regularly patrol the boundaries of their territories, making it possible to predict the movements of a particular dragonfly by observing its flight patterns: where it confronts other males, where it turns about or hovers, and where it perches. Invasion by alien males results in aggression, and aerial dog-fighting is common. Confrontations occasionally result in mid-air fights, particularly in the larger species. These scuffles involve considerable body contact and an audible clattering of wings. I once observed two male Green Darners (*Anax junius*) locked in a savage mid-air battle over a pond. Like hawks with talons locked they wrestled furiously, remarkably managing to stay airborne. As the apparent victor began to take control he forced his opponent downward into the water! Eventually the loser was able to shake off the water and the experience, and retreat.

Mating

Mating in dragonflies is intriguing because of the position the pair adopts and because it often occurs in flight. Depending on the species the male may pursue a female that has ventured into his territory or he may display or perform a simple courtship dance. With the appendages at the tip of his abdomen he firmly clasps the back of the female's head or thorax. Once this initial coupling has occurred the male, curling his abdomen, transfers a spermatophore from the tip of his abdomen to the copulatory organs on segment 2. What follows is one of the loveliest and most extraordinary reproductive behaviors in the insect world, exhibited by only one other ancient group, the Mayflies (Order Ephemeroptera). With her head or thorax still held by the male's claspers, the female grips his abdomen tightly with her claws. So secure is her grip that the male's abdomen may be scarred. Then, to receive a spermatophore from the male, the female bends her body down and forward so that the opening at the tip of her abdomen touches the copulatory organs at the base of the male's abdomen. The result is a peculiar, but graceful wheel or heart-shaped coupling of male and female dragonfly (Figure 12). The length of time the pair remains in the wheel varies with species, lasting only a few seconds in the Whitetail (*Libellula lydia*) to several hours in some Forktail damselflies (*Ischnura* sp.). The wheel is frequently assumed in mid-air and is a particularly ungainly activity in the large, heavy-bodied darners. In an interesting twist, if a male dragonfly mates with a female who has already mated, he may first scoop the previous male's spermatophore from her body with the accessory appendages at the base of his abdomen!

With mating complete, the pair will either separate or remain in tandem until ovipositing is safely accomplished. Mate-guarding protects the female from both predators and other males who may attempt to steal her away.

Figure 12. Copulating dragonflies (male above).

Migration

Migration in dragonflies is rather poorly understood but is believed to be associated with overpopulation, changes in the habitat, and perhaps weather. Although several species migrate along our coasts, the most well-known wanderer is the Green Darner, (*Anax junius*). Seasonal movement in this species is extraordinary in that spring migrants generally fly north and reproduce in Northern regions but their offspring, emerging in late summer, fly south. Massive long-distance flights have also been observed in the Globetrotter (*Pantala flavescens*), a superb flier well-known for its wanderlust. Swarms of these aptly named dragonflies have been observed at sea several hundred miles from a likely home base. Other confirmed migrants include the Tenspot (*Libellula pulchella*) and many of the big darners (Aeshnidae).

At times migratory gatherings of dragonflies may number over a million insects. On Cape Cod squadrons of migrating Green Darners have been reported from beaches along Cape Cod Bay and Monomoy Island. Upon emerging from the aquatic habitat many adult dragonflies wander from the breeding site, some traveling several miles as they feed and mature. However, large-scale migrations like those described above are limited to a relatively small number of species.

Aquatic Habitats

Because of the aquatic nature of their early lives, dragonflies and damselflies are always associated with water. In North America an incredible variety of suitable habitat exists. Dragonflies breed in clear, oxygen-rich lakes, stagnant murky pools, temporary ponds, peat bogs, cranberry bogs, swamps, brackish and fresh marshes, slow-moving streams, fast-flowing rivers and even, occasionally, swimming pools. On Cape Cod where the land was molded by glacial action and erosion, available habitats are somewhat less diverse. Although we have no great rivers like the Connecticut of western New England, we are graced with a multitude of Coastal Plain Ponds which are, despite their predominance on Cape Cod, a globally rare and threatened entity. Indeed, over 300 of these sparkling jewels dot the landscape. The nature of these glacial kettles is as diverse as the life they support. Some, like Cliff Pond in Brewster, are deep, clear, sandy lakes while others, like Shankpainter Pond in Provincetown, ooze with muck and floating *Sphagnum*. Most, however, are nearer the sandy-bottomed stage, though these too will inevitably fill in with decaying matter. Few unaltered bogs and cedar swamps remain on Cape Cod since so many natural peatlands were converted to agricultural land during the height of the cranberry industry. Nevertheless, certain Odonate species are able to colonize cranberry bogs. Coastal brackish marshes provide habitat for only a few specialized dragonfly species, since most are unable to tolerate a salty environment.

Flowing water habitats with very specially adapted Odonate species are restricted to relatively small, slow-moving streams such as Paine's Creek in Brewster and the Mashpee and Child's Rivers of the Upper Cape. These are lazy, meandering waters, narrow in width and lacking the rushing rocky currents of the big inland rivers.

The distribution and abundance of various aquatic habitats influence the types of dragonflies found in the area, since some species have very specific habitat requirements. For example, on Cape Cod the Saltmarsh Dragonfly (*Erythrodiplax berenice*) is quite widespread. It is found only in salt or brackish marshes but is widely distributed here because of the abundance of its preferred habitat. In contrast, species with strict sphagnum bog affinities, such as the charming little Bluebell (*Nannothemis bella*), are rare on the Cape, because of the scarcity of open sphagnum bogs here. As might be expected, dragonflies preferring rapidly flowing waters, such as the big river Clubtails, are also restricted here, again due to the absence of preferred habitat.

At the opposite end of the spectrum are those cosmopolitan dragonflies for whom almost any available water is a likely breeding spot. Species occurring in a wide range of habitats include the Green Darner (*Anax junius*), the Forktail damselfly, (*Ischnura verticalis*), the Blue Pirate (*Pachydiplax longipennis*), and the Damson (*Libellula incesta*).

The Coastal Plain Pond

The Coastal Plain Pond is a habitat type which is limited in occurrence to those areas situated along the Atlantic Coastal Plain. In New England this low, relatively flat seaside landscape occurs in its truest form only in Barnstable and Plymouth Counties. It reoccurs in a long strip from New Jersey southward along the coast. Cape Cod's Coastal Plain Ponds, created 12,000 years ago by melting glacial ice, range in size from tiny half-acre quagmires to great windswept lakes of several hundred acres. Many are naturally acidic and sit isolated from one another, though only a few feet apart, with no inlet or outlet of flowing water. Consequently, the surface water in these ponds represents the exposed water table, which fluctuates dramatically from season to season and year to year. In fact, the sole source of recharge for Coastal Plain Ponds is precipitation. During dry years, broad expanses of gently sloping beach are exposed around the perimeter of the ponds, and a very special complement of plants sprouts up to take advantage of the new dry land. Their seeds, having lain dormant for several years, now germinate with great urgency, assured of a sturdy foothold above water. Great sweeping stands of Plymouth Gentian (*Sabatia kennedyana*) paint the shoreline pink, and dense emergent thickets of Military Rush (*Juncus militaris*) stand tall and straight against the water. The beautiful Plymouth Gentian, rare elsewhere, is a uniquely Coastal Plain wildflower. Arrowheads (*Sagittaria* sp.), sedges (*Carex* sp.) and insectivorous Sundews (*Drosera* sp.) are also elements of the diverse Coastal Plain Pond flora, some of which are found nowhere else in the world but southeastern Massachusetts.

Although dragonflies are not linked to individual plant species as butterflies are to their food plants, the presence of typical Coastal Plain Pond vegetation can be a good indicator of which dragonfly species are present. The diversity of dragonflies at Coastal Plain Ponds is astounding, with some locales supporting 45 or more species. The spectacular Long-legged Green Darner (*Anax longipes*), the Red Bluet (*Enallagma pictum*), and the elusive Barrens Bluet (*Enallagma recurvatum*) are but a few of these Coastal Plain Pond dragonflies, true Cape Cod specialties.

Geographical and Seasonal Distribution

Cape Cod, a narrow peninsula stretching 35 miles into the sea, represents the northernmost extension of the Atlantic Coastal Plain. Enveloped by climate-moderating ocean waters, Cape Cod seasons are more temperate than those inland. Thus dragonflies with southern origins such as the Sand Dragon (*Progomphus obscurus*), and the Banded Pennant (*Celithemis monomelaena*), find Cape Cod a favorable place to live, existing here at or near the northern limits of their range. Conversely, some northern cold-water fauna including the Whiteface Skimmer (*Leucorrhinia frigida*), the Canada Blue Darner (*Aeshna canadensis*), and several Spreadwinged Damselflies (*Lestes* sp.) take advantage of our moderate climates in extending their ranges a bit southward. Indeed, Cape Cod supports a preponderance of southern species, while many truly boreal dragonflies of specialized cold-water habitats remain rare or absent here. Approximately ⅔ of the Odonate species recorded from Cape Cod are of southern origins.

The distribution of adult dragonflies varies also with the seasons. That is, emergence of a given species occurs yearly during a specific season. Spring species such as the Baskettail (*Epitheca cynosura*) and the Barrens Bluet (*Enallagma recurvatum*) often exhibit explosive mass emergences, with the majority of the population emerging from the habitat on the same day. On Cape Cod spring species emerge from warming Coastal Plain Ponds in mid-May. Many spring dragonflies are short-lived, flying for only a few brief weeks, with most of the population dying by July. Summer species such as the flashy Whitetail (*Libellula lydia*) and the vibrant Goldenwings (*Libellula auripennis*) have a longer emergence schedule spread out over several months time. These species fly throughout the months of June, July and August here. Certain summer species are most abundant very late in the season, their numbers peaking from August through September. These dragonflies, including some Blue Darners (*Aeshna* sp.) and Meadowflies (*Sympetrum* sp.), are more autumnal as adults, and these are most often the last to be seen on the wing as fall days chill into winter. Even as October closes and November begins, there are still a few hardy Meadowflies catching the very end of the sun's seasonal warmth.

Collecting and Studying Dragonflies

Rich and resplendent in color, swift and agile in flight, dragonflies present the observer with endless opportunities to study and learn. They are best observed in the field as they dart about shoreline vegetation or sail effortlessly over the water. Binoculars will bring their activities into close focus, though at times it is difficult to follow the quick turns and sudden dashes. For the collector dragonflies are a formidable challenge with their superb vision and extraordinary speed. Some species are rarely seen in collections simply because they are too speedy, too agile, or too lofty in flight to be caught.

The most efficient means of capture for adult dragonflies and damselflies is the standard insect net. Although frustrated collectors have resorted to slingshots and even shotguns for the very biggest and swiftest of the Anisoptera, these methods are not recommended and rarely yield a good specimen! I find that a net with extra wide opening and long handle, plus a little patience, are helpful for snaring the speedy high-flying darners.

Once in the net these rugged insects can be handled quite easily without damage to either dragonfly or collector. Carefully closing the wings together over the insect's thorax allows examination of its dazzling colors and huge compound eyes. After close study the dragonfly can be released unharmed to fly again over the pond. Only the teneral adults are likely to be damaged when handled.

For the Odonatologist, however, it may be necessary to collect and preserve the dragonflies as documentation of the species occurrence or to allow study of variation or anatomy. In many cases closely related species must be placed under magnification before they can be accurately identified. A dissecting microscope or handlens brings out the finest details of wing venation and the minute abdominal appendages.

Unfortunately, the grand colors of the living dragonfly fade miserably upon death. The beautiful iridescence we see in adult dragonflies is the result of light of varying wavelengths reflecting off the insect's body. In contrast, elegant reds, blues, greens and contrasting blacks and browns are pigment colors, all of which may lose much of their depth upon death of the dragonfly. Colors that are the result of pigments fade as the pigments break down, and even structural colors resulting from reflected light may be lost due to shrinkage of the dragonfly's body. Thus it is essential to keep the insect alive until it can be prepared with acetone in the lab. Adult dragonflies can be stowed safely in Glacine envelopes and kept alive in a cool dark box for several hours. Once prepared the specimens are stored in clear plastic envelopes (made especially for dragonflies) with a card of typed information. Precise locality, date and the collector's name are kept with each specimen in the collection.

The traditional method of preparing insects is to pin individual specimens with wings spread flat and body straight. However, a collection of dragonflies, which may number into the thousands, would occupy a great deal of space using this technique.

Dragonflies filed in boxes take up considerably less space than pinned specimens. Furthermore, dried specimens are fragile and long slender abdomens and legs frequently break. In pinned collections these parts may be lost or separated from their original "owner".

Odonate collecting is of vital importance to our knowledge of taxonomy, habitat requirements, behavior and even conservation. The significance of study by observation and collecting is illustrated perfectly in the case of the Barrens Bluet Damselfly (*Enallagma recurvatum*). This elusive little damsel, unknown anywhere for 32 years (1954-1986), was considered extirpated (locally extinct) in Massachusetts until 1986, when with extensive collecting it was rediscovered by the author on Cape Cod. It is now known from several pristine Coastal Plain Ponds.

Though collectors should be cautious, the removal of small numbers of dragonflies from a population is generally harmless. Far more devastating is the destruction of habitats through development, pollution or simple carelessness.

Conservation

The most serious threat to dragonflies on Cape Cod continues to be development and the spin-off pollution, trash and destruction of pond-shore communities that it causes.

Cape Cod is one of the most rapidly developing regions in Massachusetts, with the lure of the seashore drawing thousands of new residents each year and the swollen population demanding services and industries to match those of many metropolitan areas.

Conservation of plant and animal species is contingent upon the health of the habitat. An unaltered habitat in its most pristine state is a healthy, diverse habitat. Conservation efforts for dragonflies are still in their infancy, and if we are to conserve these valuable insects, we must consider them as just one element of the complex, interwoven ecosystems which they inhabit. Protection of the diversity and uniqueness of Cape Cod's Odonate fauna can come only if we protect their habitats. But before we can protect the habitats we must know where the rarities are. And so, the initial step in conservation of dragonflies is the basic field research which allows us to document populations of rare species.

Odonates in all stages of their lives are impacted by a number of activities in and around ponds, marshes, bogs and rivers. Alteration of the aquatic habitat and its surroundings, including clearing of vegetation and even stocking of fish, can be detrimental to dragonflies. In the 1950s and 1960s, reclamation efforts designed to enhance the quality of lakes for sport-fish may have had a negative impact on certain Odonates. At Flax Pond in Bourne, use of the chemical Rotenone and subsequent stocking of waters with Large-mouthed Bass and Alewives, probably reduced a fine population of the Barrens Bluet (*Enallagma recurvatum*). The creation of "pure" habitats for a few desirable animals leads to an irreparable breakdown of the elaborate, delicately balanced network of aquatic life. Pollution, whether it falls as acid in the rain or drains from well-traveled roads, ultimately upsets the health of ponds by lowering oxygen levels so vital to aquatic organisms. Fertilizers and pesticides which leach from lawns and golf courses also impact ponds by causing algal blooms and increased mortality in aquatic animals, including dragonflies.

Careless recreational activities are also punishing our Coastal Plain Ponds. Motorized off-road vehicles, thrashing through pond shore plant communities, have devastated the rich complement of rare wildflowers surrounding many ponds and churned up the protective mucky bottom sediments so vital to dragonfly nymphs. Mountains of ugly trash gradually pile up both in the ponds and on their shores.

Today several rare and endangered Odonate species are known from Cape Cod. Perhaps the most localized is the little Barrens Bluet, (*Enallagma recurvatum*). A true native of Coastal Plain Ponds, this damselfly is restricted to only four eastern states and seems to be associated with very pristine habitats. The handsome Spring Blue Darner (*Aeshna mutata*), a big boreal dragonfly, occurs only marginally in southern New England, whereas the Brotherly Clubtail (*Gomphus fraternus*) frequents the rapidly flowing rivers that are so scarce here. Cape Cod is also host to the regal Long-legged Green Darner (*Anax longipes*) and the Lateral Bluet Damselfly (*Enallagma laterale*), both designated species of Special Concern by the Massachusetts Natural Heritage and Endangered Species Program. All bear careful watching in future years as development continues to stress their fragile Coastal Plain Pond habitats.

Classification and Nomenclature

The Order Odonata is divided into two Suborders, the Anisoptera (dragonflies) and the Zygoptera (damselflies). Both of these groups are further subdivided into families. Damselflies are separated into 4 North American families, 3 of which are represented on Cape Cod. Five of the 7 North American dragonfly families are common on the Cape. The various families can be recognized with relative ease, and it is useful to become familiar with the characteristics of each. Odonate families and diagnostic characteristics are discussed in the family key and descriptions which follow.

Each individual dragonfly species is given a two-word Latin or scientific name with the generic classification first, species second.

For the Long-legged Green Darner (*Anax longipes*), *Anax* is the genus, *longipes* the species. Scientists always use the Latin name when referring to a species and, as you will see, there are good reasons to do so. Nevertheless, common names can be useful, as charming and alluring as the insects themselves, and in many cases they are good descriptive terms. In general, common names for families and genera are more standardized and widely used than those at the species level where there is often no English name at all. From here things get confusing. Among Odonatologists, usage of common names differs. Some scientists dream up their own descriptive names, which after all is the way most originate! As a result there may be two or three common names in the literature for the same species. For example, *Pachydiplax longipennis*, a Common Skimmer (Family Libellulidae) which I have long known as the Blue Pirate, is also called the Blue Dasher; and *Libellula pulchella* may be labeled either Tenspot or Twelvespot depending on which spots you count! *Epitheca princeps*, in the Family Corduliidae (sometimes Green-eyed Skimmers, sometimes Emeralds), is called either the Royal Skimmer, the Prince Baskettail, or simply the Prince!

Further complicating matters is the fact that each family and genus has its own English name (sometimes two), and these are not always the same. So, the Little Bluebell (or perhaps the Elfin Skimmer) belongs to the Dwarf Skimmer genus *Nannothemis* in the Skimmer Family Libellulidae. However, in the case of the monster darner *Epiaeschna heros*, both the genus *Epiaeschna* and the species *E. heros* are called Hero Darners . . . or sometimes Swamp Darners! With such confusion, one can see why Odonatologists ultimately depend upon the Latin names.

Until recently there was no consistent, sensible system of common names for the Odonata, resulting in the kind of confusion illustrated above. Laborious searches of the literature sometimes produced accepted names, but quite often did not. In 1986, however, a scheme was created by two Odonatologists, Dennis Paulson and Sidney Dunkle, which is far more complete and practical than anything previous. While most of the English names in this book were adopted from Paulson and Dunkle's *English Names for North American Odonata* (in press), old habits die hard and I find it difficult to drop some of the more historic names with which I have grown familiar.

Glossary

Note: The terms listed here are defined as they are used in the text. For anatomical parts please refers to the illustrations on pages 15-20.

Abdomen: the long, tail-like segment of the dragonfly's body, or the third segment of an insect's body.

Anal appendages: extensions at the tip of the Odonate abdomen, the shape of which may be diagnostic in males. Males use these claspers to hold the female during mating and oviposition. Male dragonflies have three appendages, damselflies four.

Anisoptera: "unequal wings," referring to the different size of the fore- and hind-wings of dragonflies; the sub-order, or taxonomic group, comprising dragonflies.

Antennae: two hair-like sense organs on the head of an insect, much reduced in dragonflies.

Antenodal: before the nodus, usually referring to the crossveins of the wing which occur between the wing-base and the nodus.

Apex: at the tip or end of.

Arculus: a crossvein between the wing base and the triangles of fore- and hind-wings.

Auricles: flaps on the second segments of many male dragonflies.

Basal: at the base or beginning of.

Caudal: toward the tail or rear.

Caudal lamellae: the tail-like gills of a damselfly nymph, of which there are usually three.

Compound eyes: the eyes of insects, spiders, crabs, etc., which contain many individual lenses (see **ommatidia**).

Costa: the thickened vein which forms the leading edge of the wing.

Crepuscular: active at twilight.

Dorsal: back or top-side of an insect.

Ecdysis: molting or shedding of the exoskeleton, which takes place during growth and emergence in Odonates.

Elongate: lengthened, long.

Emergence: the transition between the larval (nymph) stage and the adult stage, when the dragonfly nymph leaves the water and becomes a flying adult.

Emergent vegetation: aquatic plants which are rooted on the pond bottom but protrude above the water's surface.

Exoskeleton: the hard, protective covering of an insect's body; an outer skeleton.

Exuvia: the cast or shed exoskeletons of the Odonate nymph.

Flavescent: tinted with amber or gold.

Hyaline: transparent.

Inferior: below; usually referring to the abdominal appendages.

Labium: the lower lip; a large, moveable, hinged structure for seizing prey in the dragonfly nymph.

Labrum: the upper lip.

Lateral: side.

Mate-guarding: a behavior associate with many Odonates, in which the male guards his mate to ensure that she is successful in ovipositing.

Metamorphosis: the development of insects, which progresses through a series of stages from egg to adult.

Molt: to shed the hard covering on the body (exoskeleton); part of the growth process.

Nocturnal: active at night.

Nodus: the notch or indentation on the front edge of the Odonate wing.

Nymph: the dragonfly larva.

Ocelli: simple eyes of adult insects, of which there are usually three.

Odonata: the order or taxonomic group of insects containing dragonflies and damselflies.

Ommatidia: individual facets of the compound eyes of insects, which number into the thousands in Odonates.

Oviposit: to deposit or lay eggs.

Ovipositor: the egg-laying apparatus on the tip of the abdomen of female insects.

Paleozoic: the geologic time period which extended from about one billion to 215 million years ago.

Post-ocular spots: spots of color on top of the head of many species of damselflies, the shape, size and color of which are characteristic of each species.

Prothorax: the neck or collar-like segment just behind the head of the dragonfly, to which the first pair of legs is attached.

Pruinose: covered with a bluish-white powdery bloom; a color change that takes place in many adult dragonfly species.

Pubescent: covered with short downy hairs.

Rectal gills: the internal gills of dragonfly nymphs.

Spermatophore: a packet containing sperm, produced by male insects and transferred to the female during mating.

Spiracles: breathing holes on the insect's body.

Stigma: a small, square or rectangular cell near the tip of each dragonfly wing, usually colored.

Superior: top; usually referring to the abdominal appendage.

Synthorax: the second and larger section of the dragonfly thorax, which holds the two pairs of wings and the hind two pairs of legs.

Tandem: a pre- and post-mating position adopted by many Odonata in which the male holds on to the back of the female's head with claspers at the tip of his abdomen.

Teneral: a newly emerged adult dragonfly, with soft bodies and pale colors.

Thorax: the middle segment of an insect's body, located behind the head, to which the wings and legs are attached.

Trachea: tubes which carry air to all parts of an insect's body.

Vulvar plate: two small plates on the underside of the female abdomen, part of the ovipositor.

Zygoptera: "similar wings," referring to the similar size of the fore- and hind- wings of damselflies; the sub-order, or taxonomic group, comprising damselflies.

A Family Key for Cape Cod Odonata

Front and hind wings of equal size, similar shape; when insect is perched, wings are folded up over its body; male with 4 anal appendages . . . . Zygoptera (damselflies), p. 32

Front and hind wings of different size and shape, the hind wings broadest; wings held flat like those of an airplane; anal appendages of male usually 3 . . . . . Anisoptera (dragonflies), p. 45

Partial Key for Cape Cod Zygoptera

1. Larger species, with metallic colors; wings broad, often fully or partially dark-colored; wings without narrow stalk at base, densely veined . . . . Calopterygidae (Broad-winged Damselflies), p. 32

1a. Usually smaller species, some with metallic colors; wings clear, with narrow basal stalk 2

2. When at rest, damselfly holds wings partially spread; males with long anal appendages; stigma of wing more than twice as long as wide . . . . Lestidae (Spread-winged Damselflies), p. 33

2a. When at rest, wings folded together over back; male appendages generally shorter, often difficult to see without magnification; stigmas no more than twice as long as wide . . . . Coenagrionidae (Narrow-winged Damselflies), p. 34

Partial Key for Cape Cod Anisoptera

1. Eyes widely separated on top of head, not touching; end of abdomen usually swollen; adults often perch on or near the ground . . . . Gomphidae (Clubtails), p. 49

1a. Eyes touching for some distance on top of head; abdomen rarely with noticeable swelling at tip 2

2. Very large (3+ inches), dark insects with blue, green or brown markings; eyes touch on top of head, and are in contact for considerable distance; two antenodal crossveins of wings thickened (Figure 9); adults rarely seen at rest . . . . Aeshnidae (Darners), p. 45

2a. No thickened antenodal crossveins; triangle of hind wing closer to arculus than that of forewing 3

3. Large, brown or black dragonflies with a single yellow stripe around the thorax, most obvious from the side; anal loop of hind wing short, wide and rounded (Figure 9); eyes in contact on top of head for a short distance . . . . Macromiidae (Cruisers), p. 56

3a. Anal loop of hind wing elongate, often shaped like a foot, comprising 2 rows of cells (Figure 8) 4

4. Medium-sized, slender and darkly colored dragonflies, often with metallic color; anal loop of hind wing long, somewhat clubbed at end but not distinctly foot-shaped . . . . Corduliidae (Emeralds), p. 52

4a. Wings often, but not always, with spots or patterns of color; thorax and abdomen lacking metallic color; anal loop of hind wing foot-shaped with a well-developed "toe" portion (Figure 8) . . . . Libellulidae (Common Skimmers), p. 58

How to Use a Taxonomic Key

Using a taxonomic key is at first a bit overwhelming, but you will find it a valuable skill to learn. Armed with the correct key and a little basic knowledge, the identity of almost any plant or animal is quite literally, at your fingertips. The Odonate keys provided here are partial keys; that is, they apply only to those groups of dragonflies which are found on Cape Cod. The key will lead you to the identity of the insect as far as the family level. Providing the reader with an Odonate key to the species level and the knowledge required to successfully use such a key would be a monumental task.

Let's practice by taking the Wandering Globetrotter (Figure 13, Plate 8) through the key as follows:

A. Note the difference in size and shape of front and hind wings, and the resting position of the wings (flat like airplane wings) —proceed to Anisoptera (dragonfly) key.

B. The eyes are touching on top of the head and the tip of the abdomen is not swollen (choice 1a—proceed to 2).

C. Antenodal crossveins not thickened and the hindwing triangle is closer to the arculus than in the forewing (see Figure 9, page 18 for diagram) (choice 2a—proceed to 3).

D. Note the very long anal loop with well-developed foot (choice 3a—proceed to 4).

E. Note the lack of metallic color and very distinct foot-shaped anal loop with well-developed toe (choice 4a) . . . thus the insect is a Common Skimmer (Family Libellulidae).

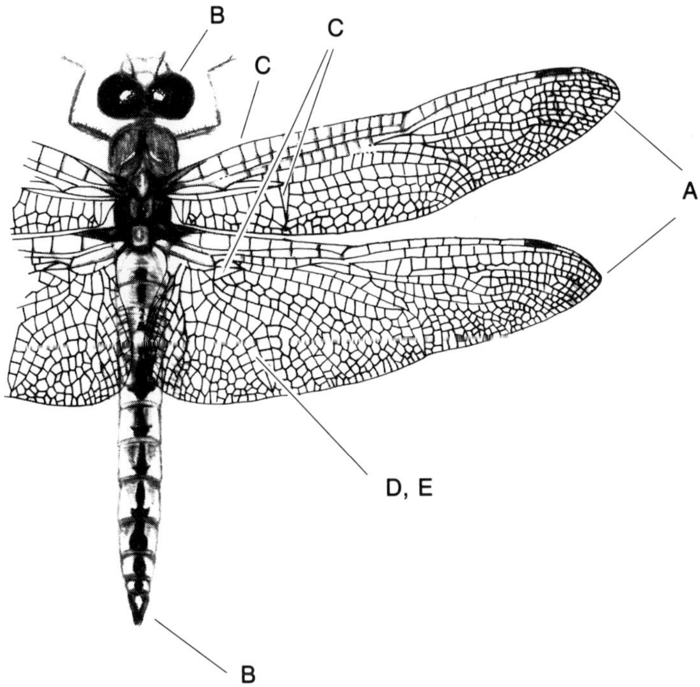

Figure 13. **Wandering Globetrotter** (*Pantala flavescens*).

Color Plates

The damselflies and dragonflies illustrated in the following plates are all males, except where noted. Since they are not necessarily life-size, actual ranges of body length are given in the captions on the facing pages. The captions contain only brief descriptions; more detailed descriptions referred to in the text should be consulted for positive species identification. Anal appendages, in particular, though not shown in closeup in the plates, are often diagnostic of a species and should be studied closely. The heads of damselflies shown in lateral view in Plates 1 and 2 are turned 90 degrees to better illustrate the post-ocular spot patterns.

PLATE 1

EBONY JEWELWING *Calopteryx maculata* . . . . p. 32
45 mm. (1 7/8 in.) Shining green body; black wings. Streams and rivers. Summer.

AMBER-WINGED SPREADWING *Lestes eurinus* . . . . p. 33
45–50 mm. (2 in.) Honey-colored wings held partly spread; metallic green body. Grassy ponds. Early summer.

SLENDER SPREADWING *Lestes rectangularis* . . . . p. 33
45–50 mm. (2 in.) Very long, slender abdomen; aqua shoulder stripes. Shady swamps, ponds, brooks. Summer.

SWAMP SPREADWING *Lestes vigilax* . . . . p. 34
43–48 mm. (1¾ in.) Iridescent green; pale blue-tipped abdomen. Weedy ponds. Summer.

VARIEGATED DAMSEL *Chromagrion conditum* . . . . p. 36
33–36 mm. (1 3/8 in.) Blue and black; bright yellow spot on side. Rivers, brooks, pools, boggy ponds. Spring to mid-summer.

EASTERN RED DAMSEL *Amphiagrion saucium* . . . . p. 35
23-25 mm. (7/8 -1 in.) Small; abdomen brick-red, black-tipped. Streams, bogs. Early summer.

SPHAGNUM SPRITE *Nehalennia gracilis* . . . . p. 44
26-28 mm. (1 in.) Tiny; iridescent green body. Bogs, boggy ponds. Early to mid-summer.

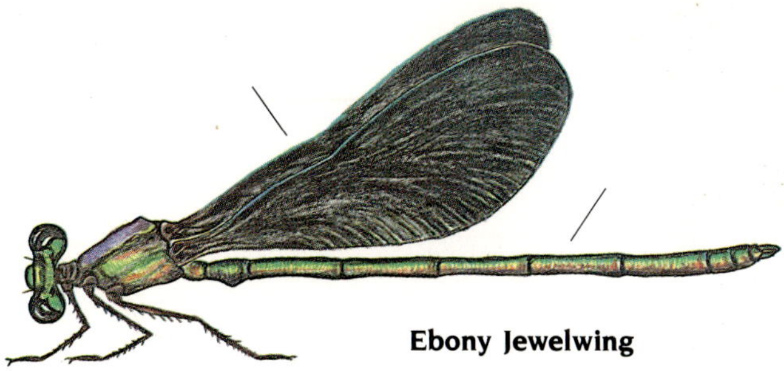

Ebony Jewelwing

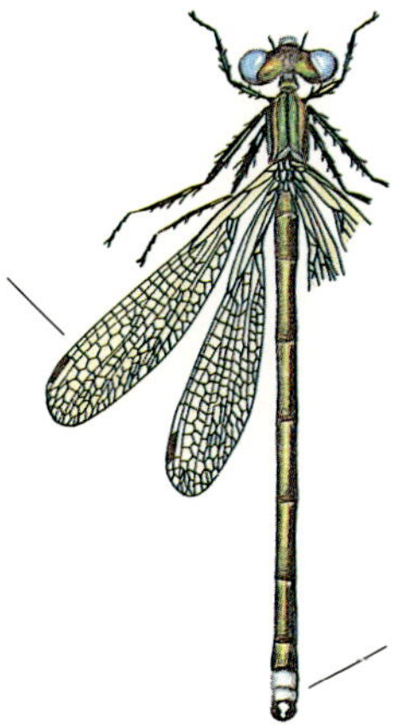

Amber-winged Spreadwing

Slender Spreadwing

Swamp Spreadwing

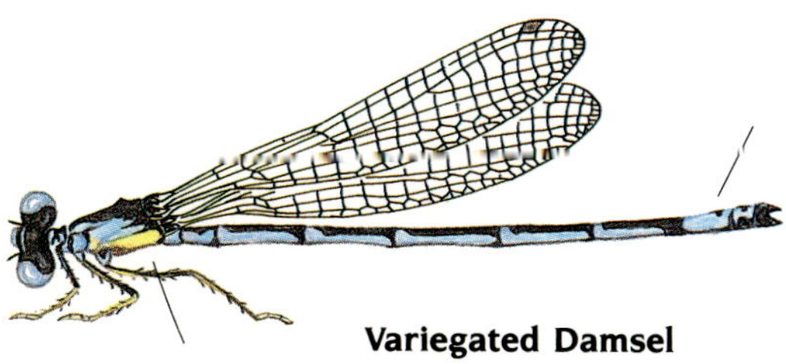

Variegated Damsel

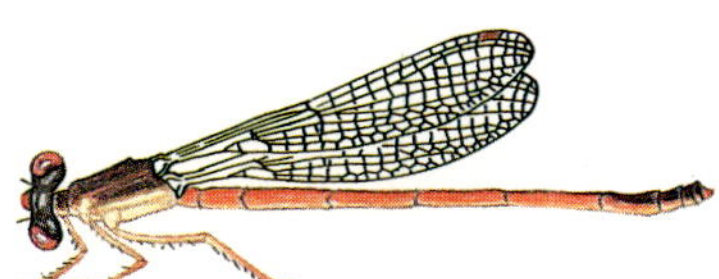

Eastern Red Damsel

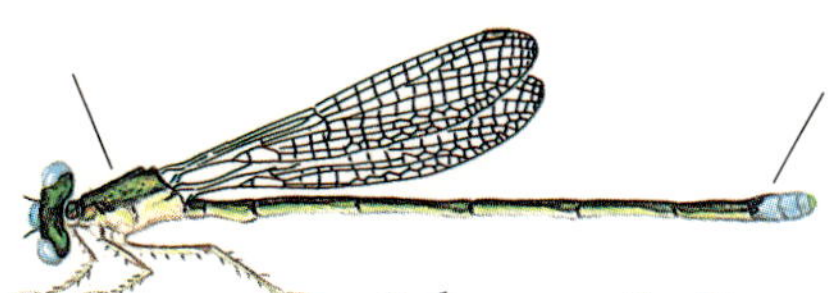

Sphagnum Sprite

PLATE 2

COMMON FORKTAIL *Ischnura verticalis* .... p. 43

25-30 mm. (1—1 ¼ in.) Green and black striped thorax; black abdomen with blue tip. Marshy ponds, lakes and streams. Spring to early autumn.

CIVIL BLUET *Enallagma civile* .... p. 37

Blue and black banded abdomen. Sandy lakeshores. Late spring and summer.

RED BLUET *Enallagma pictum* .... p. 40

29-33 mm. (1 ⅛—1 ¼ in.) Bright red face and thorax; shiny black abdomen. Weedy pondshores (especially water lilies). Summer.

ORANGE BLUET *Enallagma signatum* .... p. 41

31-35 mm. (1 ¼—1 ⅜ in.) Orange and black thorax; slender black abdomen with orange spot near tip. Lakes, sandy ponds. Summer.

EVENING BLUET *Enallagma vesperum* .... p. 42

30-34 mm. (1 ¼ in.) Yellow and black head and thorax; black abdomen with pale blue spot near tip. Crepuscular. Ponds. Summer.

VIOLET DANCER *Argia fumapennis* .... p. 36

31-35 mm. (1 ¼—1 ⅜ in.) Violet or purple and black. Streams, rivers, ponds and lakes. Summer.

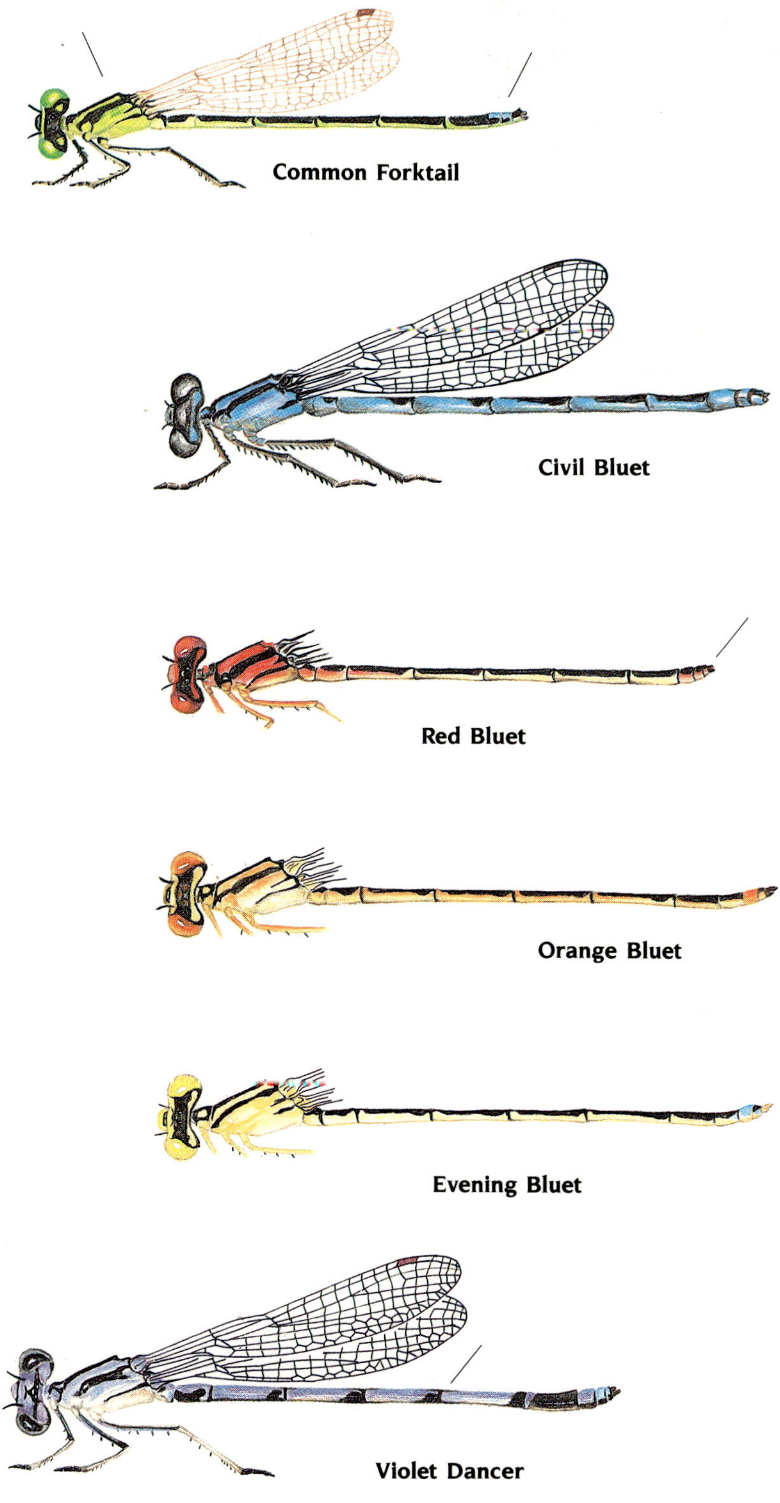
Common Forktail
Civil Bluet
Red Bluet
Orange Bluet
Evening Bluet
Violet Dancer

PLATE 3

BLACK-TIPPED DARNER *Aeshna tuberculifera* . . . . p. 46

76–82 mm. (3–3¼ in.) Large size; mosaic blue and black pattern; distinctive thoracic striping. Grassy ponds. Late summer, autumn.

SPOTTED BLUE DARNER *Aeshna clepsydra* . . . . p. 46

66–70 mm. (2¾ in.) Slender, small Aeshnid; sides of thorax spotted. Grassy ponds. Autumn.

SHADOW DARNER *Aeshna umbrosa* . . . . p. 47

68–75 mm. (3 in.) Male appendages with spine on tip; thorax stripes narrow. Often flies late afternoon and evening. Streams, pools, ditches. Late summer, autumn.

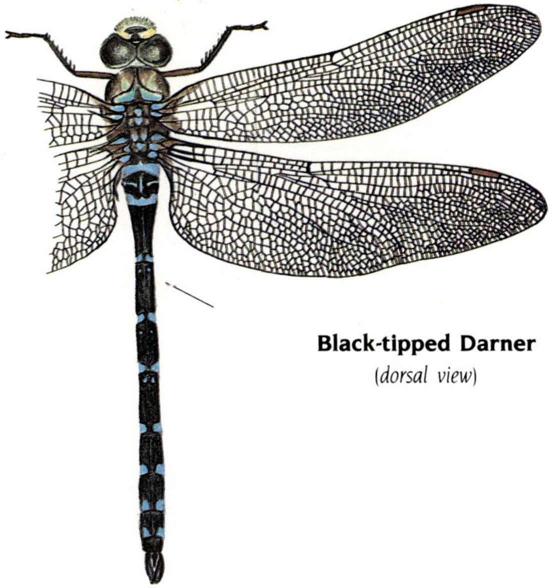

Black-tipped Darner
(dorsal view)

Black-tipped Darner
(lateral view)

Spotted Blue Darner

Shadow Darner

PLATE 4

COMMON GREEN DARNER *Anax junius* . . . . p. 47

70-80 mm. (3—3¼ in.) Large size; blue abdomen; bulls-eye pattern on forehead. Ponds. Early summer to fall.

LONG-LEGGED GREEN DARNER *Anax longipes* . . . . p. 48

75-87 mm. (3—3½ in.) Large size; red abdomen; long brick-red legs. Grassy ponds. Summer.

LANCET CLUBTAIL *Gomphus exilis* . . . . p. 51

38-42 mm. (1½—1¾ in.) Small ground-percher; slight flaring at tip of abdomen. Sandy ponds, lakes, rivers. Spring to mid-summer.

SAND DRAGON *Progomphus obscurus* . . . . p. 51

50-53 mm. (2 in.) Moderate flaring of abdomen; short legs; ground percher. Sandy ponds and lakes. Summer.

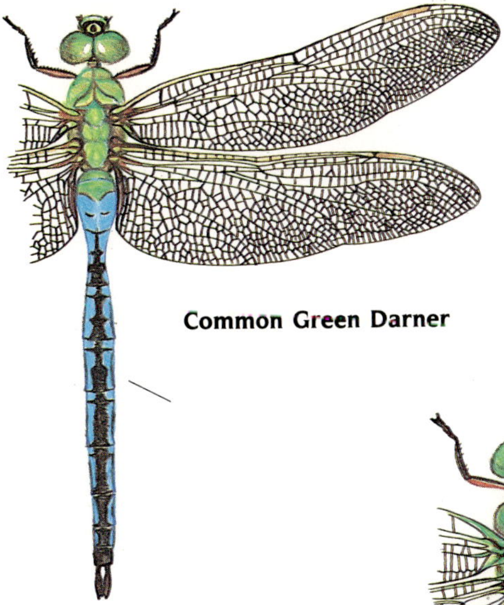

Common Green Darner

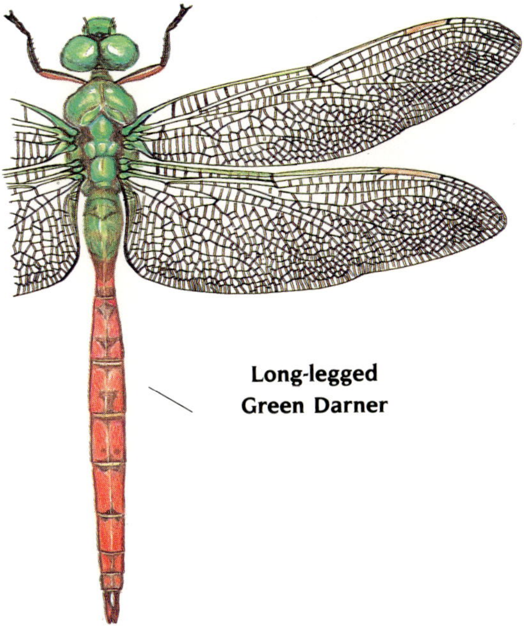

Long-legged
Green Darner

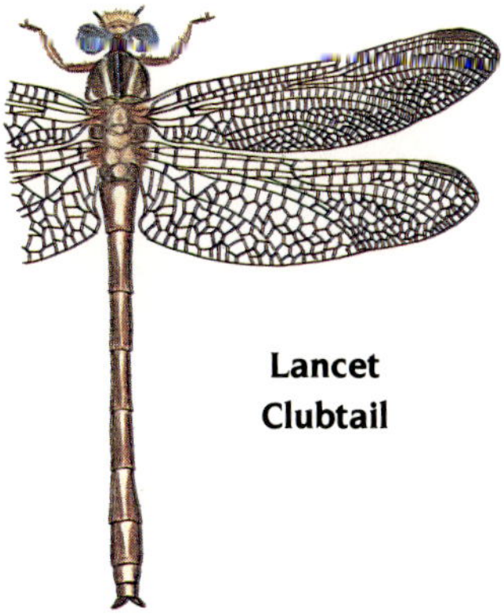

Lancet
Clubtail

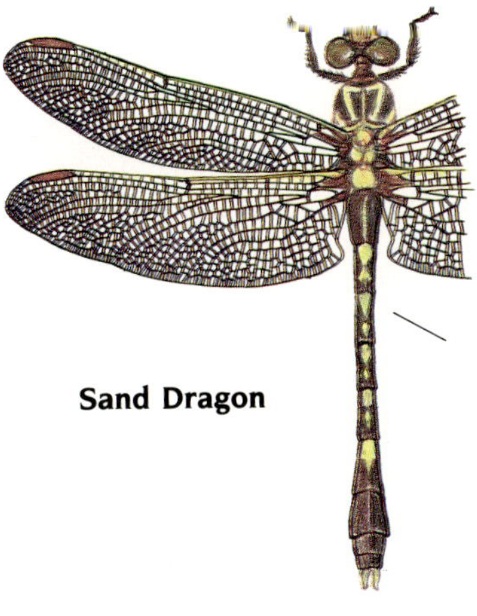

Sand Dragon

PLATE 5

SWIFT RIVER CRUISER *Macromia illinoiensis* . . . . p. 57

64–67 mm. (2½–2¾ in.) Single belt around thorax; spotted abdomen. Rivers and sandy ponds. Summer.

STREAM CRUISER *Didymops transversa* . . . . p. 57

52–57 mm. (2¼ in.) Belted thorax; spotted abdomen. Rivers and sandy ponds. Summer.

COMMON BASKETTAIL *Epitheca cynosura* . . . . p. 54

37–41 mm. (1⅝ in.) Often feeds in large groups. Lakes, ponds, swamps, marshes and streams. Early to mid-summer.

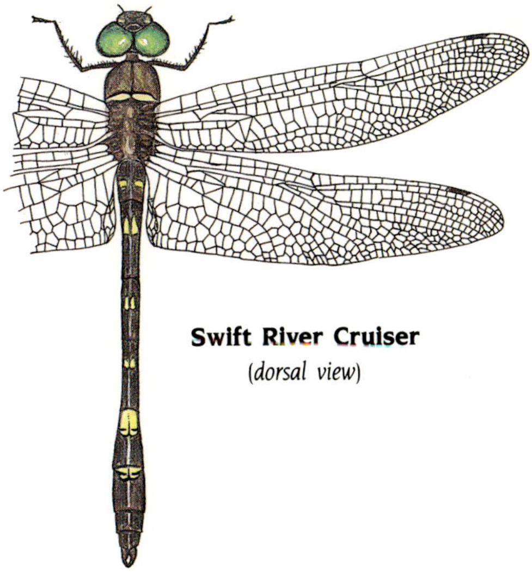

Swift River Cruiser
(dorsal view)

Swift River Cruiser
(lateral view)

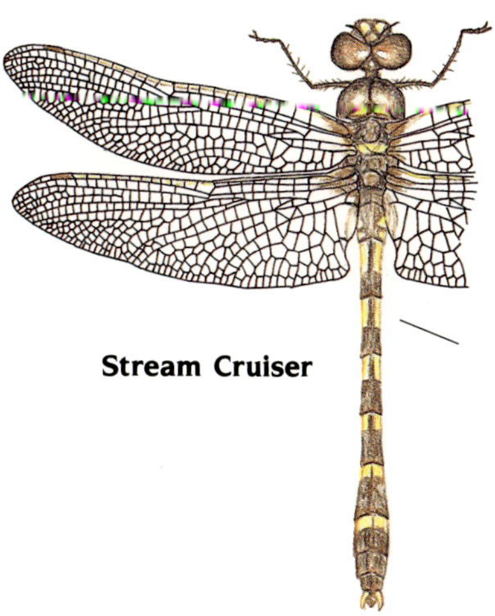

Stream Cruiser

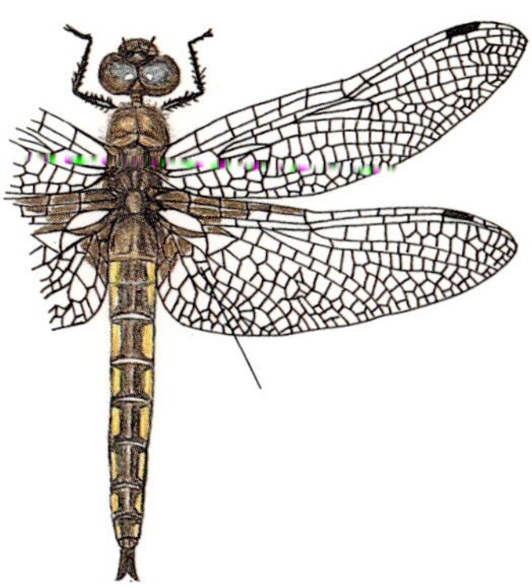

Common Baskettail

PLATE 6

MARTHA'S PENNANT *Celithemis martha* . . . . p. 59

29–32 mm. (1¼ in.) Dark color; single wing spot. Ponds. Summer.

ELISA SKIMMER or CALICO PENNANT *Celithemis elisa* . . . . p. 58

29–32 mm. (1¼ in.) Wing spotted; body mostly red. Lakes, ponds, streams, marshes. Summer.

WINGS: **HALLOWEEN PENNANT** *Celithemis eponina* . . . . p. 59

BANDED PENNANT *Celithemis monomelaena* . . . . p. 59

BLUE PIRATE *Pachydiplax longipennis* . . . . p. 68

36–40 mm. (1½ in.) Weedy ponds. Summer.

GREEN JACKET *Erythemis simplicicollis* . . . . p. 60

39–44 mm. (1½–1¾ in.) Resembles female when teneral. Sandy lakes, ponds, streams. Summer.

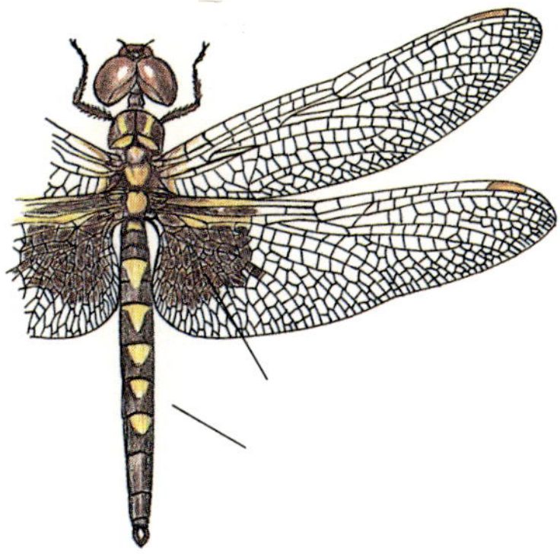

Martha's Pennant

Elisa Skimmer or Calico Pennant

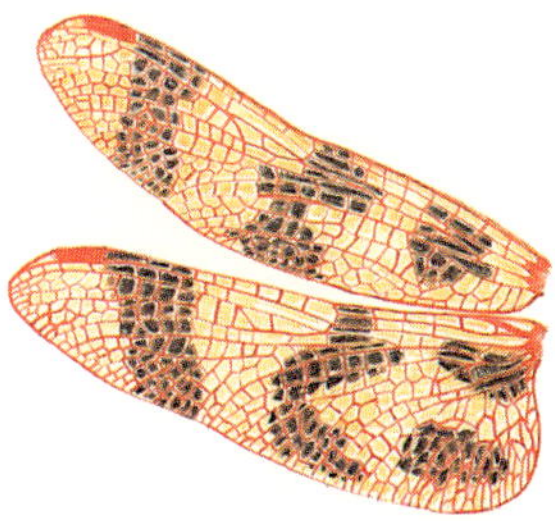

Halloween Pennant

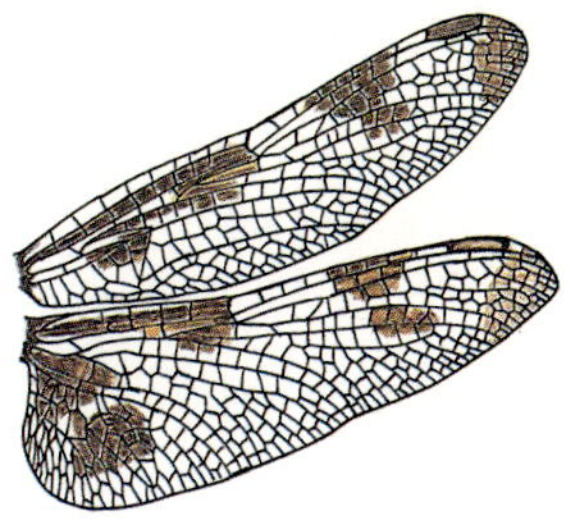

Banded Pennant

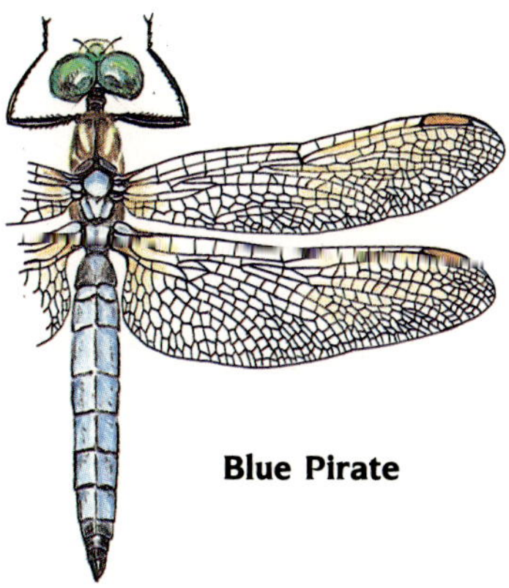

Blue Pirate

Green Jacket (*male*)

Green Jacket (*female*)

PLATE 7

DAMSON SKIMMER *Libellula incesta* . . . . p. 64

48–52 mm. (2 in.) Dark body; clear wings. Ponds and lakes. Summer.

GOLDENWINGS *Libellula auripennis* . . . . p. 63

51–56 mm. (2–2¼ in.) Wings pumpkin colored. Grassy Ponds. Summer.

TENSPOT *Libellula pulchella* . . . . p. 65

50–53 mm. (2 in.) Black and white spotted wings. Ponds. Summer.

WHITETAIL *Libellula lydia* . . . . p. 65

42–45 mm. (1¾ in.) White abdomen; wings with dark large band; ground percher. Ponds, rivers, marshes. Late spring through late summer.

CORPORAL SKIMMER *Libellula exusta* . . . . p. 64

34–40 mm. (1⅜–1⅝ in.) Pale abdomen; shoulder stripes; ground percher. Ponds. Spring.

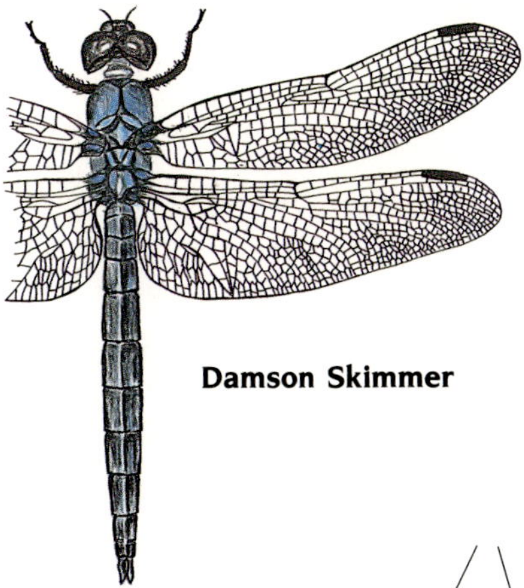

Damson Skimmer

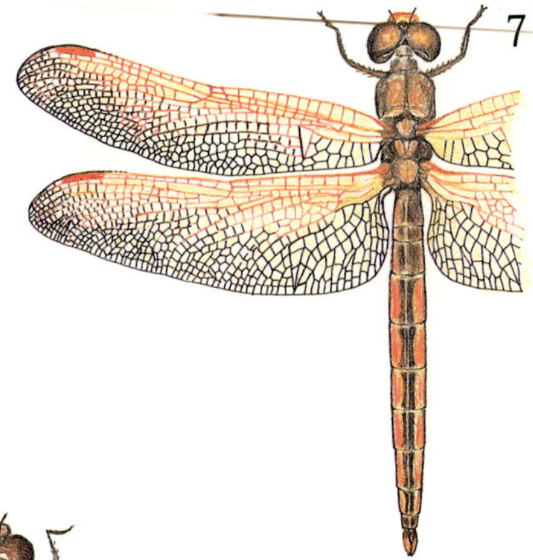

Goldenwings

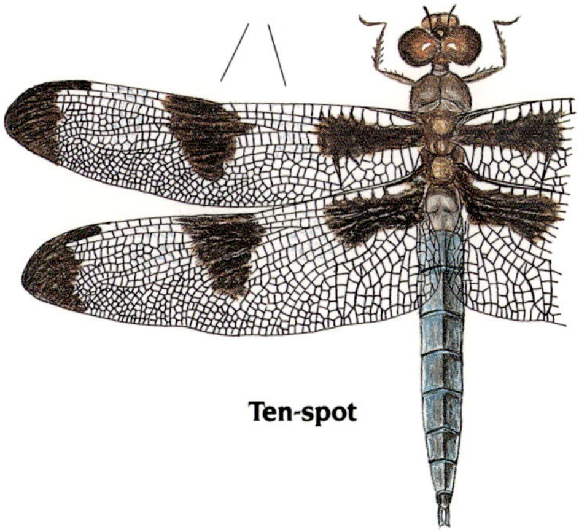

Ten-spot

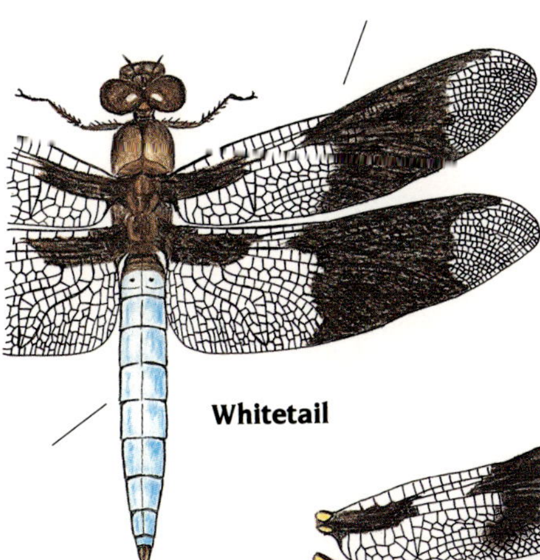

Whitetail

Corporal Skimmer

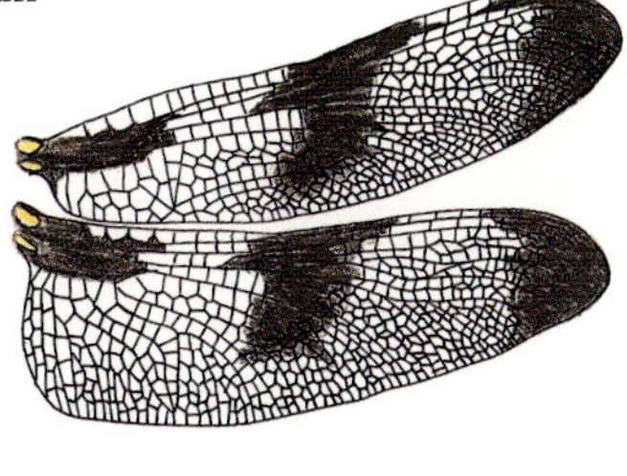

Whitetail
(*female wings*)

PLATE 8

RUBY MEADOWFLY *Sympetrum rubicundulum* .... p. 70

35–37 mm. ($1\frac{3}{8}$ in.) Red body. Grassy ponds. Summer to autumn.

YELLOW-LEGGED MEADOWFLY *Sympetrum vicinum* .... p. 71

32 mm. (1¼ in.) Pale legs. Ponds. Summer to autumn.

RED SADDLEBAGS *Tramea carolina* .... p. 71

48-52 mm. (2 in.) Broad wings; large basal spot; short abdomen. Grassy ponds. Summer.

WANDERING GLOBETROTTER *Pantala flavescens* .... p. 68

50 mm. (2 in.) Long broad wings; often feeds in large groups. Ponds. Summer.

AMBERWINGS *Perithemis tenera* .... p. 69

21–24 mm. ($\frac{7}{8}$–1 in.) Small size; uniform color; tinted wings. Ponds. Summer.

Ruby Meadowfly

Yellow-legged Meadowfly

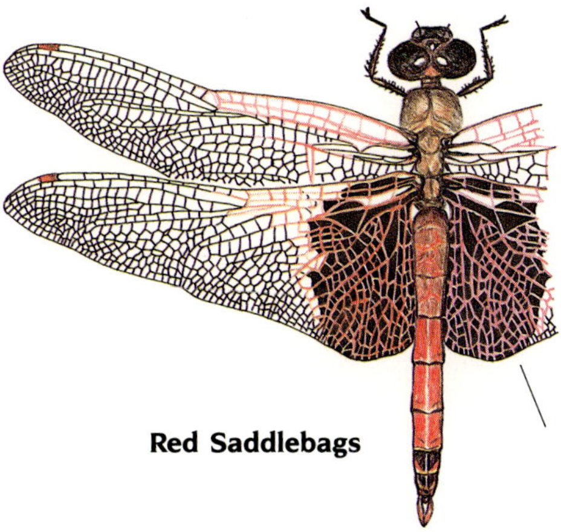

Red Saddlebags

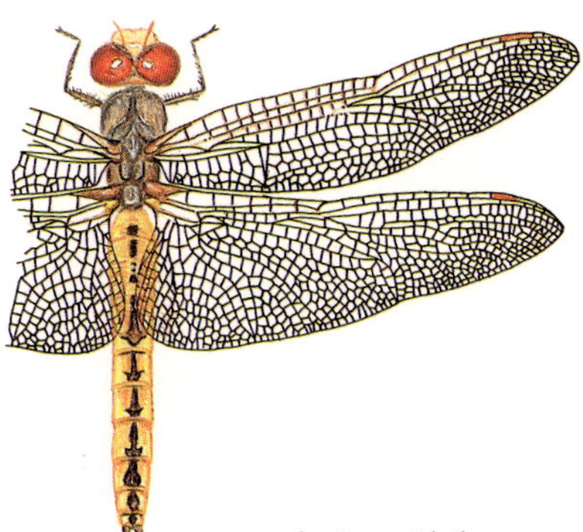

Wandering Globetrotter

Amberwings

Species Descriptions

In the following pages I have described in detail 61 (approximately 2/3) of Cape Cod's Odonate species. Most of those I have chosen to introduce to you are common here or elsewhere in their range, or are good representatives of a particular family or group. I have also chosen those species which I thought the reader would be most likely to "bump into" in his or her Cape Cod wanderings. For larger genera, I have included a brief introduction to the genus.

Measurements given are taken directly from specimens collected in Massachusetts and represent the insect's length from head to tip of the abdominal appendages. Because most of the dragonflies we see on our Cape Cod ponds are males, I have concentrated descriptions and illustrations on this sex. I hope that the descriptions will enable the reader to identify the dragonflies he or she encounters, if not to the species level, then at least to the correct genus.

The Damselflies

(Suborder Zygoptera)

Broad-winged Damselflies

(Family Calopterygidae)

Broad-winged Damselflies are large, strikingly colored insects which inhabit rivers and streams. Adults are metallic-bodied with dark, often opaque wings. This is a small family, composed of just two North American genera and 8 species. We have 2 representatives of the group in our area: the very common Ebony Jewelwing (*Calopteryx maculata*) and the beautiful American Rubyspot (*Hetaerina americana*), which is known only from a historic record (pre-1920). The latter species, with brilliant crimson wing marks, is a southern damselfly recorded from much of the United States. Though there is no reason to doubt its existence on the Cape, the Rubyspot frequents those rapidly flowing streams that are so scarce here.

Black-winged Damsels (Genus *Calopteryx*)

All 5 North American species of Black-winged Damsels, also known as Jewelwings, possess deep iridescent green bodies and dark wing marks. The amount of black on the wings varies with species, from the entire wing surface to a dipped-in-ink effect at the tips. On Cape Cod only the Ebony Jewelwing (*Calopteryx maculata*) is known.

EBONY JEWELWING (*Calopteryx maculata*) **Plate 1**

Description: One of the larger and more exquisitely colored of the damselflies, the Ebony Jewelwing measures to about 45 mm. (1⅞ in.) Adults of both sexes boast a shining metallic green head, thorax and abdomen. The male has dusky coal black wings which, rather surprisingly, lack stigmas. The wings of the female are brown with dark tips and white stigmas. Leg spines are exceedingly long relative to other damselflies. The abdominal appendages of the male are long and slender, black in color.
Flight Period: June through September.
Range: Eastern U. S., south to Florida and west to Texas. On Cape Cod, Barnstable, Mashpee, Harwich and Brewster, but probably throughout on most fresh running brooks and rivers.

Comments: This lovely iridescent damselfly is commonly encountered along shaded rivers and streams where it can be extremely numerous, with hundreds of individuals congregating in very small areas. The Ebony Jewelwing exhibits a fluttery, almost butterfly-like flight. Adults are almost always seen in close proximity to running water, where they light on vegetation overhanging stream beds like so many little black bows tied gaily to the tips of the branches. Females oviposit alone in a variety of aquatic vegetation, while the male remains close to ward off potential danger or intruding males.

Spread-winged Damselflies
(Family Lestidae)

The Spread-winged Damselflies include some of the largest North American Zygoptera, adults of which may measure more than 2 inches in length. They can be instantly recognized by the peculiar manner in which they hold their wings. Resting adults sit poised with wings partially spread, hence the common name Spreadwings. There are just two genera in the family, one of which occurs on Cape Cod.

Spread-winged Damselflies (Genus *Lestes*)

Adult Lestes are active insects, found perched in shoreline shrubbery or emergent vegetation where their rather dull coloration is a highly effective camouflage mechanism. Females have conspicuous, well-developed ovipositors for slicing into plants. Overall body color and abdominal appendage shape are important criteria in species identification, though as a whole this is a distinctive group of insects. Nymphs are long-bodied and sprawling with very slender legs. They frequent the weedy places of quiet marshy ponds and lakes. Many of the 9 species recorded from the Cape are common, but 2 of these are represented only by historic records dating to 1917.

AMBER-WINGED SPREADWING (*Lestes eurinus*) **Plate 1**

Description: This is a large, stout-bodied damselfly, measuring to 45-50 mm. (2 in.) The head, thorax and abdomen are metallic green above and pale buff below. Older individuals acquire a frosty blue pruinosity on top of the thorax and on the last 2 abdominal segments. Adults can be easily recognized by their large size, shining metallic green color and the honey-colored flavescence of the wings. Male anal appendages are distinctive, the superiors very long and curved inward from above, the inferiors short, blunt and much less than ¼ the length of the superiors.
Flight Period: June and July.
Range: Southern Canada, New England, New York and Pennsylvania west to Michigan and Missouri, south to North Carolina. Amber-winged Spreadwings are common on Cape Cod, occurring from Bourne and Falmouth to the Cape tip.
Comments: This large, conspicuous damselfly is found in grassy Coastal Plain Ponds where it perches on rushes and other shoreline plants, a foot or so above the water. Large numbers of mated pairs may be seen, the males retaining their hold on the female even through oviposition.

SLENDER SPREADWING (*Lestes rectangularis*) **Plate 1**

Description: This handsome damselfly is almost ungainly with its absurdly long abdomen and short wings. Adults measure 45-50 mm. (2 in.), the abdomen 40-42 mm. of the length in larger specimens. The face is pale aqua to nearly blue in some, the top of the head black and the eyes a lovely ocean blue. The dark, coppery brown thorax shines with a faint bronzy light. A very fine aqua stripe runs from the head to the wing-base and there are 2 wide aqua shoulder stripes as well. The thorax is buffy yellow on the sides and below, and the legs are pale. The exceeding long, slender abdomen is dark bronze above with a greenish light, pale yellowish on the sides and below. Females, though similar in color, are shorter and stockier.

Flight Period: July and August.

Range: Nova Scotia and New England, west to North Dakota, south to Oklahoma, the Gulf States and Florida. The species is widespread on Cape Cod, from Bourne and Falmouth to Provincetown.

Comments: I have seen this species in a variety of habitats including Atlantic White Cedar Swamps, mucky ponds and brooks, but it seems to prefer shaded situations. Adults, when flushed from pond shore vegetation, fly directly into surrounding woodlands or shrubbery where they hang beneath leaves and twigs, perfectly concealed.

SWAMP SPREADWING (*Lestes vigilax*) **Plate 1**

Description: This is another metallic green species, a bit smaller than L. *eurinus* at 43–48 mm. (1¾ in.). The head and face are very dark, but the labrum (upper lip) is a pretty pale robin's egg blue. Both the thorax and segments 1 through 8 of the abdomen are deep metallic green above and yellow or pale below. The sides of the thorax are marked with black stripes and spots. End abdominal segments, black in teneral individuals, become a striking pruinose white with age. The male's superior appendages are long, gently curved, the inferiors very thin and straight, about ¾ the length of the superiors.

Flight Period: June through September.

Range: Nova Scotia and Maine west to Minnesota and Texas, south to the Florida Everglades. *Lestes vigilax* is known from every Cape Cod town.

Comments: This is the most abundant Spreadwing on Cape Cod, its numbers incredibly high almost everywhere it is found. Like others of its genus, L. *vigilax* prefers the shelter of pond shore vegetation, perching in emergent rushes and sedges well above the water. I have collected it at a wide variety of pond habitats from mucky quagmires to deeper sandy kettles.

Narrow-winged Damselflies
(Family Coenagrionidae)

Narrow-winged damselflies are small creatures, most only an inch or so in length. They inhabit a variety of freshwater systems, though most prefer quiet still waters where nymphs are concealed in underwater vegetation. Adults are usually spotted or banded with blue, green and black, and most lack the metallic lustre of the previous families. The wings, narrowed to a slender basal stalk, have little noticeable color. Adults are timid by nature with relatively weak flight, and are most easy to capture with an insect net. They remain close to the habitat, seldom wandering from the immediate surroundings except during the short maturation period or when weather drives them to shelter. Nymphs are shorter and stouter than either the Agrionidae or Lestidae, with a triangular labium and prominent branching veins on the caudal gills.

Mating may last for several hours, after which the male retains his grasp on the female. During oviposition, while the female clings to a supporting plant stem, the male stands nearly vertical to the water's surface, the tip of his abdomen locked tightly behind the female's head. Eggs are laid inside the stems and leaves of floating, emergent or submergent vegetation.

Identification of most groups in this family can be relatively simple. A handful of useful characteristics—including color pattern, wing venation and the shape of the anal appendages will help in determining the identity of various species. A good hand lens is standard field equipment for magnification of appendages, and I often depend upon the dissecting microscope for a final species determination.

The Coenagrionidae is our largest, most diverse damselfly family, represented in North America by 15 genera and nearly 100 species. On Cape Cod we have 6 genera and at least 24 species. Their names are as captivating as the insects themselves: the Bluets (genus *Enallagma*) with vivid blue and black hues and the Dancers (genus *Argia*) with purple and violet body colors; the tiny Green Sprites (*Nehalennia*) and Forktails (*Ischnura*); and the striking Red Damsels (*Amphiagrion*). In the following descriptions I will focus on species which are commonly encountered on Cape Cod or are unique to our area. Unfortunately, it is not possible to include every species, but the beginning Odonatologist should be content with a general familiarity with the six locally occurring genera. Perhaps the more adventuresome will delve further into the subtle differences between even the most similar of species.

Red Damsels (Genus *Amphiagrion*)

The genus *Amphiagrion* contains only two North American species, one of which is found in our area.

EASTERN RED DAMSEL (*Amphiagrion saucium*) **Plate 1**

Description: This is a small, richly colored damsel, the male painted with bold reds and blacks. Adults range from 23–25 mm. ($^7/_8$–1 in.), the female usually a bit larger than the male. Males, with their short thick abdomens, appear quite stout-bodied. The head is black above without postocular spots, while the face and mouthparts are brick-colored. The top of the thorax is black, becoming pale beneath and the abdomen is bright orange red for about ⅔ its length, after which it is coal black on top. Females are brown or reddish brown.

Flight Period: June and July.

Range: Nova Scotia and New England south to Georgia, west to Illinois and Minnesota. On Cape Cod, I have collected this species in Mashpee and Wellfleet.

Comments: The Eastern Red Damsel is not particularly common on Cape Cod. It is restricted to small streams or slow moving rivers here, though elsewhere in its range it is found in boggy habitats and muddy quagmires. I observed it in Wellfleet in a stagnant, grassy part of the Herring River where the stream was sunny, narrow and shallow and vegetation plentiful. Adults flew near the ground, staying within the shelter of vegetation.

The Dancers (Genus *Argia*)

We have just two of the North American species of Dancers on Cape Cod, one of which I have not yet seen, but for which several historic records exist. Dancers are usually associated with gentle stony streams and rivers, but our common Cape species, *Argia fumipennis*, is found in a variety of pond and lake habitats as well. The English name Dancer may be derived from the restless, flighty behavior of these damselflies. Males rest only briefly and bat their wings together in an agitated manner upon lighting. During oviposition the male grasps the female with his appendages in the typical tandem position, standing vertical to the water as she backs down a plant stem. In some species, the female or sometimes the pair, remain completely submerged for more than 30 minutes. Though a seemingly risky behavior, male and female usually emerge unharmed.

VIOLET DANCER (*Argia fumipennis*) **Plate 2**

Description: An unmistakable, medium-sized *Argia* (31–35 mm. or 1¼–1 3/8 in.) with striking violet color at maturity. In males, the face and head are purple, with very large postocular spots sometimes connected by a narrow purple band. The thorax too is purple, with a black stripe on top and on each shoulder. The abdomen is a vibrant purple-violet alternating with narrow black bands, with a lovely powdery blue tip. The chunkier females are dull purple or brown, with black bands and streaks on the abdomen. The larger, darker A. *moesta* (40+ mm.) shows a cinnamon and black thorax and uniform black abdomen, but becomes pale pruinose white with age.

Flight Period: June through early September.

Range: New England, Nova Scotia and New Brunswick west to Minnesota, Oklahoma and Arizona, south to Florida. On Cape Cod, A. *fumipennis* is common from Bourne to Provincetown.

Comments: Violet Dancers choose open sunny resting places on or near the ground. I find them perched on rocks, logs and woodland paths, a habit distinctive of this genus. They are wary, nervous and quick to fly when approached. Pairs are conspicuous and often seen on ponds, where they oviposit in wet logs and aquatic plants.

Variegated Damsels (Genus *Chromagrion*)

This North American genus contains only one species and it is indeed quite rare on Cape Cod.

VARIEGATED DAMSEL (*Chromagrion conditum*) **Plate 1**

Description: A big, slender, handsomely colored damselfly measuring 33–36 mm. (1 3/8 in.) The color pattern of adults is blue and black, but the species is easily recognized by the contrasting primrose yellow on the sides of the thorax. The face is pale blue, the top of the head black. The thoracic pattern, black on top, sky blue and sunny yellow on the sides, is a conspicuous identification feature. The abdomen is black above, each segment with a narrow blue ring at its base. Segments 8 and 9 are pale blue with tiny paired black dots on top. Segment 10 is black. Females are similarly colored but the abdomen is entirely black.

Flight Period: May through July.

Range: New England to Wisconsin, south to Missouri and Georgia. It is known from Barnstable, Yarmouth and Mashpee.

Comments: I have collected only a few individuals of this species from local waters. Adult males were found patrolling low over the ditches of a cranberry bog, an environment that would seem not particularly well-suited to aquatic life. Elsewhere, *Chromagrion conditum* frequents rivers and brooks, the edges of ponds near streams and sometimes spring-fed pools or very boggy pond shores.

Bluets (Genus *Enallagma*)

These are the familiar blue and black damselflies of Cape pond shores. They are small and slender with clear wings. While most have handsome cerulean blue and black coloration, a few species are bright orange or red. Postocular spots are present behind the eyes and are variable in shape and size. Species can be identified by a combination of features including color pattern and wing venation, but the most accurate means of determining closely related species is examination of the minute appendages on the tip of the abdomen. The genus is enormous, with about 45 North American species, of which 17 are recorded from Cape Cod.

BOG BLUET (*Enallagma aspersum*) **Figure 14**

Description: Relatively large for a Bluet, E. *aspersum* measures 30–35 mm. (1¼–1⅜ in.) The face is blue, the top of the head black with very large triangular blue postocular spots. The thorax is blue with a wide black band on top and one narrower band on each side, the slender abdomen blue and black as follows: the first few segments mostly blue; segments 4-6 entirely black above; the first half of 7 black, the second half blue; 8-10 blue. The most significant identification marks are the large triangular spots behind the eyes, the mostly black abdomen, bicolored on segment 7 and the appendages. Females are similar to males, except segment 7 is mostly blue with a narrow black stripe on top widening rearward. Segments 8-10 are black in the female, 8 with two blue spots.

Flight Period: Early June through August, most abundant in July.

Range: Maine west through southern Quebec to Montana, south through Oklahoma and South Carolina. *Enallagma aspersum* occurs from Falmouth to Provincetown on Cape Cod.

Comments: This pretty blue damselfly is a resident of classic Coastal Plain Ponds with shallow grassy or boggy shorelines. It is often found in ponds where there is very little water or where vegetation is dense throughout and the bottom quite mucky. I have also collected the Bog Bluet at swamps carpeted with *Sphagnum*. Adults are strong fliers, but generally stay close to the water. They are not averse to venturing out over open water, away from the protective jungles of shoreline plants. Males actively drive off intruders and appear quite aggressive.

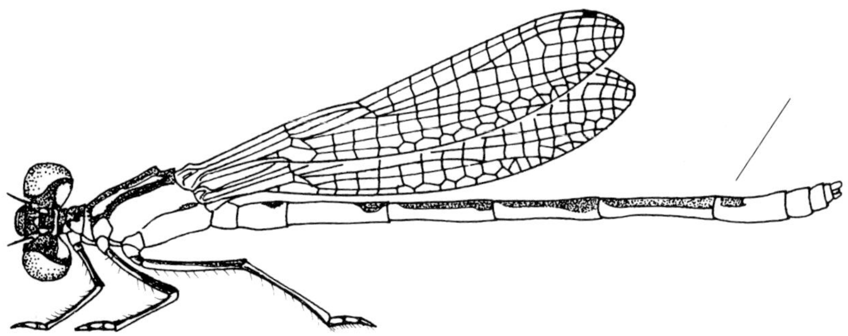

Figure 14. **Bog Bluet** (E. *aspersum*)

CIVIL BLUET (*Enallagma civile*) **Plate 2**

Description: A slender, mostly blue *Enallagma* measuring 31–37 mm. (1¼–1⅜ in.). The postocular spots of the head are pale blue, small and narrow, often connected by a thin blue line. The thorax, though patterned like others of this genus, becomes a deep royal blue in mature males. The abdomen of males is alternately banded with black and bright turquoise blue, the blue bands 4 times as broad as the black on segments 3-5, the black on 6 lengthened on top to a pointed spot covering half the segment. Segment 7 is almost entirely black above, 8 and 9 blue, and 10 blue with a black mark on top. The female's abdomen is black on top, buff to light blue below.

Flight Period: Late May through mid-August.

Range: Nova Scotia, Maine, west through Saskatchewan south to California, Utah, Arizona, Texas and Florida. Also from Mexico, Central America and northern South America. On Cape Cod, E. *civile* is widespread from Bourne and Falmouth to Truro.

Comments: The Civil Bluet, though very much like Doubleday's Bluet, is only occasionally found in company with it. On Cape Cod, E. *civile* shows a marked preference for deeper, clear, sandy lake shores with little mud or vegetation.

COMMON or NORTHERN BLUET (*Enallagma cyathigerum*) **Figure 15**

Description: A big, husky *Enallagma* with vivid, electric blue predominating on the thorax and abdomen and very large postocular spots which nearly touch the eyes. Adults measure 28-34 mm. (1⅛–1¼ in.). The thorax, intense blue in fully mature males, has the usual black stripes on top and on the shoulders. Segments 1–5 of the male's abdomen are mostly blue with narrow black bands, 6 and 7 primarily black. Segments 8 and 9 are entirely blue, 10 blue with a black spot on top. The female is much like the male, but the abdomen is black above, the blue reduced to a narrow ring on each segment. This species is virtually identical to E. *vernale*, which also occurs in our area, and the two are now considered conspecific by many entomologists.

Flight Period: Early spring: mid-May through July.

Range: Extends far north to southern Alaska and Eurasia, southeast to Newfoundland, through New England to New York and west to southern California, Nevada and Utah. On Cape Cod, E. *cyathigerum* is widespread from Falmouth to Provincetown.

Comments: The Northern Bluet, an abundant early-flying spring species, prefers the quiet shores of ponds, bogs and occasionally swamps. It is a strong flier, readily moving out over unprotected waters where it skims along close to the surface in the typical Bluet fashion.

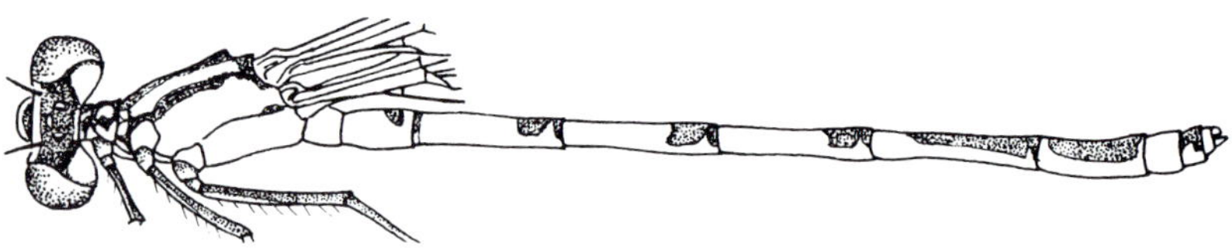

Figure 15. **Common or Northern Bluet** (E. *cyathigerum*)

DOUBLEDAY'S BLUET (*Enallagma doubledayi*) **Not illustrated**

Description: Another predominantly blue species, very much like E. *civile* and most effectively separated from it by the form of the male's abdominal appendages, and the habitat. Adults may appear subtly smaller 31-35 mm. (1¼-1⅜ in.) and a bit chunkier. Females are extremely difficult to separate from those of E. *civile*, but may be identified by the company they keep!

Flight Period: Late May through mid-August.

Range: Coastal Massachusetts to Florida and Louisiana. On Cape Cod, it is widespread and abundant from the Canal to Provincetown.

Comments: Doubleday's Bluet is one of the most abundant summer damselflies on Cape Cod. It shows a particular affinity throughout its coastal range for sandy-bottomed ponds which may have thick muddy deposits and dense shoreline vegetation. Adults are strong fliers which range widely over the open pond waters.

SKIMMING BLUET (*Enallagma geminatum*) **Figure 16**

Description: Our smallest *Enallagma* at 24-29 mm. (7/8 - 1 1/8 in.). E. *geminatum* is easily identified by its size, coloration and behavior. The head and thorax are marked like other *Enallagmas*, the head with comma-shaped postocular spots. The abdomen is almost entirely black except for segments 8 and 9 which are pale blue, and the basal segments which are blue on the sides. The distinctive second abdominal segment shows a diamond-shaped black spot on top and a narrow wavy band low on the side. This unmistakable identification mark is present in both sexes.

Flight Period: June through August.

Range: Eastern U. S. from Texas and Michigan east to Maine, south to Florida. E. *geminatum* is abundant Cape-wide.

Comments: This diminutive little damsel prefers still ponds, often muddy with dense aquatic vegetation, others more open but still with ample shoreline vegetation. Adults rest flat on floating water lily leaves and are difficult to capture without swamping the net. They skim slowly and gracefully over the water, within millimeters of its surface.

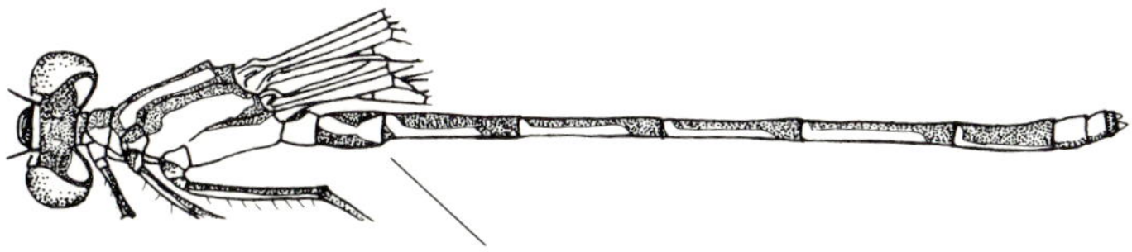

Figure 16. **Skimming Bluet** (E. *geminatum*)

LATERAL BLUET (*Enallagma laterale*) **Figure 17**

Description: A small, pale blue and black banded *Enallagma* measuring 25-29 mm. (1 - 1 1/8 in.). The Lateral Bluet is named for the conspicuous but variable black mark on the sides of the 8th abdominal segment. The face and head are blue and black with rather small, comma-shaped postocular spots. The thorax is blue with the usual black stripe on top and on the shoulders. The abdomen is alternately banded with blue and black, blue predominant on the first five segments, black becoming dominant rearward. Segments 8 and 9 are blue, 10 black. Segment 8 shows a distinctive lateral black mark. The combination of the mostly blue abdomen, black lateral band on segment 8 and the shape of the appendages will identify males. Females are black above, pale buff below.

Flight Period: An early spring species with brief flight period, late May through the end of June.

Range: Northeastern: southern Maine south through New Hampshire, eastern Massachusetts, Rhode Island and Connecticut to northern New Jersey and Pennsylvania. E. *laterale* is widespread on Cape Cod, occurring from Falmouth and Bourne to Truro.

Comments: The Lateral Bluet, though restricted in range, is thriving in eastern Massachusetts (particularly Plymouth and Barnstable Counties). It is known from more than 30 Massachusetts locations. In Rhode Island, *Enallagma laterale* is the dominant spring Odonate, outnumbering virtually every other species flying. Here its habitat of choice is the classic Coastal Plain Pond, usually heavily vegetated with emergent rushes, the beautiful pink Plymouth Gentians and purple Pickerelweeds, and often with mucky or boggy edges. It has also been recorded at sphagnum bogs.

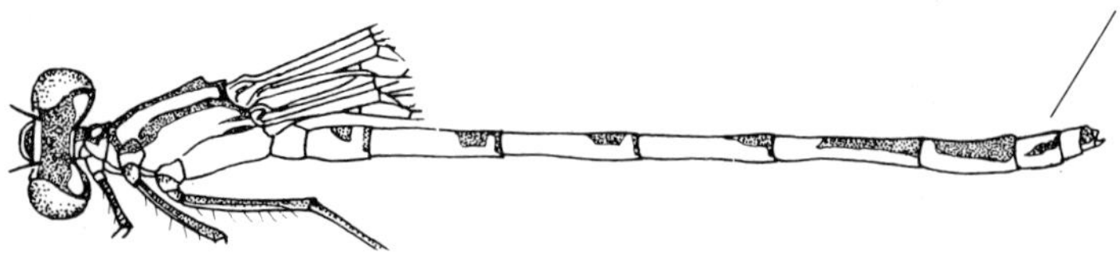

Figure 17. **Lateral Bluet** (E. *laterale*)

LITTLE BLUET (*Enallagma minusculum*) **Figure 18**

Description: Another small damsel whose specific name *minusculum* undoubtedly arose from its stature. Most adults range from 25-27 mm. (1 in.). The face and head are blue and black, the postocular spots small and comma-shaped. The blue thorax has the typical black banding but is different from others of this genus in the pretty purplish tint acquired by mature males. The abdomen is alternately banded with deep blue and black, the black bands wider than the blue, giving the insect an overall dark appearance. Segments 8 and 9 are entirely blue. The male's appendages are minute but distinctively shaped. Females are black above, buff below.

Flight Period: June through August, with most activity in June.

Range: Restricted to coastal Northeast: Maine, Nova Scotia and New Brunswick south through New Hampshire, Massachusetts, Rhode Island, Connecticut, New York to North Carolina. It is a dominant and abundant species Cape-wide.

Comments: This petite little Bluet thrives along well-vegetated Coastal Plain Pond shores, where males patrol emergent rushes and grasses, perching along exposed plant stems. Populations may be huge, numbering into the thousands, even quite early in June. At Horseleech Pond in Truro, 95% of all damselflies seen were males of this species, even as early as June 11. Almost all were heavy with parasitic water mites.

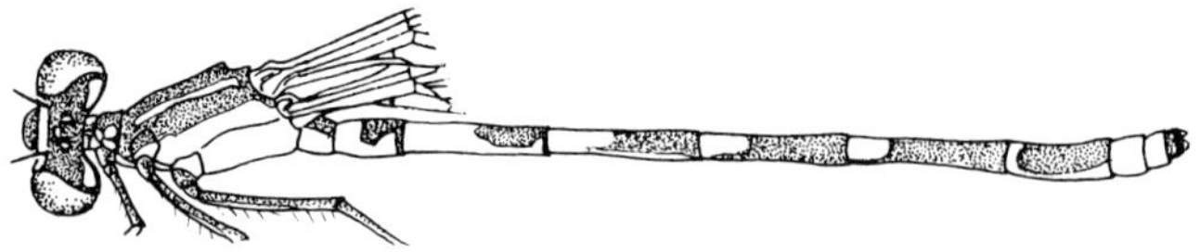

Figure 18. **Little Bluet** (E. *minusculum*)

RED BLUET (*Enallagma pictum*) **Plate 2**

Description: A striking, unmistakable *Enallagma* with bright red and black color. Adults are slender-bodied, measuring 29-33 mm. (1⅛-1¼ in.). Where most *Enallagmas* are blue, this species is orange-red or crimson red and thus quite conspicuous. The face is red, the top of the head black with a narrow red band behind the eyes instead of the usual 2 spots. The thorax is red, striped with black in the usual manner. The abdomen, entirely black above through segment 10, becomes reddish or buff below. Females are black above, yellow beneath.

Flight Period: Late June through mid-August.

Range: Very restricted: Massachusetts, New York and New Jersey. On Cape Cod, it flourishes at many locations from Falmouth to Wellfleet and probably on to Provincetown.
Comments: This beautiful *Enallagma*, though very localized in occurrence, finds the Coastal Plain Ponds of Cape Cod ideal habitat. It is known from at least a dozen Upper Cape kettles, but is never as abundant as other Bluets. Adults frequent shorelines where water lilies abound, using the large floating leaves as resting spots.

BARRENS BLUET (*Enallagma recurvatum*) **Figure 19**

Description: The Barrens Bluet is closely related to the Lateral Bluet, but quite different in appearance. It is a small, deep blue and black species measuring 26-29 mm. (1 - 1⅛ in.). The head and thorax are similar to E. *laterale* but the black bands on the abdomen are wider than the blue, which gives E. *recurvatum* a much darker appearance. From above, the alternating black bands become progressively wider to segment 7, which is entirely black. Segments 8 and 9 are blue, except a variable black dash on the side of 8 which is absent in some individuals and as conspicuous as that of E. *laterale* in others. Final determination of this species should depend upon examination of the male's abdominal appendages: the superiors are short and abruptly upturned, giving the insect its specific name, *recurvatum* (see Figure 11). The female's thorax is patterned like the male's, but the abdomen is shining black above with a greenish lustre, buff below. The only noticeable blue on her abdomen is a small basal spot on segment 8.
Flight Period: Early and brief, late May throughout June, with some populations flying into July.
Range: Very restricted, eastern Massachusetts, Rhode Island, coastal New York and New Jersey. On Cape Cod, it is widespread from Bourne and Falmouth to Truro.
Comments: A true Coastal Plain species, the Barrens Bluet inhabits shallow pond shores where vegetation is plentiful. It is not found in deep, clear sandy kettle ponds, but prefers those with muddy deposits and often with dense stands of rushes and Plymouth Gentian. Females oviposit a few inches above the water in the stems of emergent rushes (often *Juncus militaris*), with the males in tandem. These are aggressive little damselflies: I have seen males attack tenerals, mated pairs, single males of their own and other species, and even dead individuals. The Barrens Bluet is frequently found with the Lateral Bluet and the Little Bluet. Despite the restricted nature of its range it is abundant throughout June at pristine Coastal Plain Ponds.

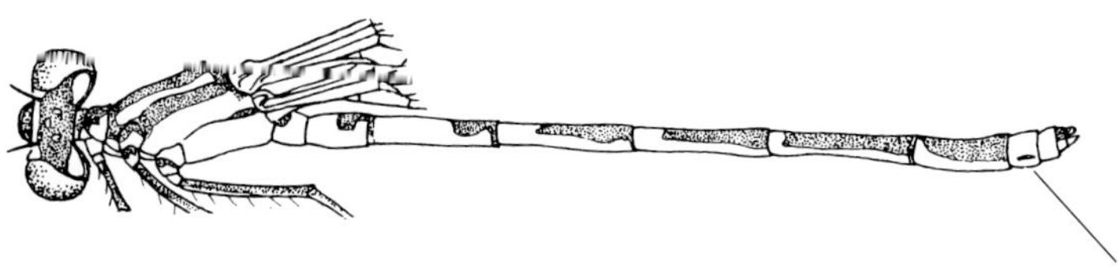

Figure 19. **Barrens Bluet** (E. *recurvatum*)

ORANGE BLUET (*Enallagma signatum*) **Plate 2**

Description: Another unusually-colored Bluet, somewhat like E. *pictum* but pale orange and black. Adults measure 31-35 mm. (1¼ - 1⅜ in.) and boast very long slender abdomens. The face is orange, the head black on top with a thin ochre band behind the eyes. The thorax, though patterned like other *Enallagmas*, is orange instead of blue. The abdomen is black above, yellow on the sides, and segment 9 is orange. Females resemble the males, but are pale yellow or buff instead of orange. The abdomen is jet black to segment 10, which is a pretty pale blue.

Flight Period: June through August.

Range: Eastern: New England west through Canada to Minnesota, south to Texas and Florida. On Cape Cod, E. *signatum* ranges from Bourne and Falmouth to Truro.

Comments: This brightly colored, slender-bodied Bluet frequents a wide variety of Coastal Plain Ponds from large sandy lakes to shallow mucky pond shores and occasionally streams. It prefers areas with floating water lilies, though males may move far from shore when chasing rivals. Females are generally shy and retiring, seen on the ponds only when mating or found perched in surrounding vegetation.

EVENING BLUET (*Enallagma vesperum*) **Plate 2**

Description: A slender, bright yellow and black species, 30-34 mm. (1¼ in.) in length with a contrasting spot of blue at the tip of the abdomen. Both the head and thorax are yellow and black, the thorax similar to other *Enallagmas* but with a very thin black shoulder stripe. The abdomen is black on top, yellow on the sides and below, while segment 9 is entirely blue. Females are similar to males, but the abdomen is shining black to segments 9 and 10, which are yellow.

Flight Period: June through August.

Range: Maine, Nova Scotia and New Brunswick west to Wisconsin, south to Texas and Florida. Here, I have collected E. *vesperum* from Bourne to Truro.

Comments: *Enallagma vesperum* is an unusual damselfly in its crepuscular or evening-active habits, and its species name reflects this tendency. However, I sometimes find adults active mid-afternoon and catch them sneaking about in the brush during late morning. They remain in the shelter of emergent plants when on the pond, but wander into surrounding woodlands during daylight hours.

Fork-tailed Damselflies (Genus *Ischnura*)

The Forktails are small, common damselflies with bright green and black thoracic color and shining black abdomens. The common name is derived from the minute forked projection on the tip of the male's abdomen. Some species are polymorphic: adults may be of several different color phases and their color changes noticeably with age. Forktails are found quite early in the spring and fly into September. They are numerous along the weedy shores of ponds and lakes, and are also seen on streams and rivers. Of the 15 North American species, 5 are known from Cape Cod.

SPEARTAIL (*Ischnura hastata*) **Figure 20**

Previously of the genus *Anomalagrion*, the Speartail was recently incorporated into the above group.

Description: A tiny slight damselfly easily recognized by its yellow and black color, minute size (23-26 mm. or ⅞-1 in.) and the prominent 2-pronged spine on segment 10 of the male's abdomen for which the species is named. The orientation of the stigmas are peculiar, with those of the forewing red in color, roundish in shape and separated from the wing margin by a short connecting vein. The head is metallic black with minute round postocular spots. The thorax, shining bronzy-black above and pale lemon on the sides, bears yellow legs and unusually short wings. The abdomen is prettily marked with black over mustard yellow: segments 1 and 2 are black above, 4-6 with dark arrow-shaped marks pointing rearward, 7 mostly black and 8-10 yellow. The complexity of the abdominal color pattern is quite remarkable up close. Females are orange and black, but of a different pattern than the males.

Flight Period: Late June through August.

Range: Southern Maine west through southern Ontario to N. Dakota, south through Florida, Cuba, Louisiana and Mexico to Venezuela. The Speartail is known from just a few Lower Cape locations.

Comments: This timid little sprite is our smallest Zygoptera, and as such may be easily overlooked. It prefers wet areas with thick rushes and very little water. I have never seen it on open pond waters, but always in the expanses of shoreline vegetation thriving in the wake of dropping summer water levels. Adults stay very low in these spike rushes and grasses, their gaily colored yellow bodies surprisingly well-camouflaged.

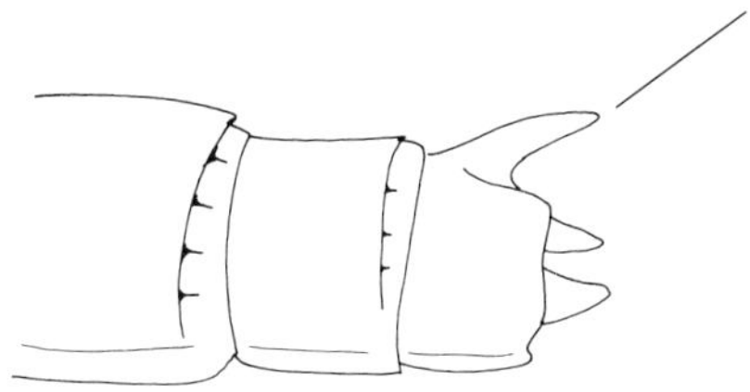

Figure 20. **Speartail** *(I. hastata), detail of segment 10 of male's abdomen showing two-pronged spine.*

FRAGILE FORKTAIL *(Ischnura posita)* **Figure 21**

Description: A delicate little Forktail, quite different from others in the genus and very easily recognized with just a glance at the thorax. The dark, slender-bodied adults measure 24-28 mm. (1 - 1⅛ in.). Males are primarily metallic-black, with yellow face and small greenish postocular spots. The thorax is jet black above with lime green shoulder stripes, each one distinctively shaped like an exclamation mark. Each segment of the glossy black abdomen has a narrow yellow ring at its base. Teneral females resemble males but become bluish grey at maturity.

Flight Period: Early May through early September.

Range: Eastern Nova Scotia and Maine south to Florida, west to Texas and the Great Plains. I. *posita* is abundant Cape-wide.

Comments: This dainty little Forktail is frequently found in the company of its close relative *Ischnura verticalis*, but is usually less numerous. It flies sluggishly through shoreline vegetation, very low to the water. Once spotted, it is easily identified by the conspicuous green exclamation points on its thorax.

COMMON FORKTAIL *(Ischnura verticalis)* **Figure 21, Plate 2**

Description: A smartly colored damselfly measuring only 25-30 mm. (1 - 1¼ in.). The tiny round postocular spots are bright green against the dark head, while the black thorax shows two brilliant lime green shoulder stripes. The slender abdomen is shining black above with a greenish lustre. Segments 8 and 9 are sky blue above with variable black marks on the sides. Females appear in several color phases, often changing dramatically as they mature. Some tenerals closely resemble the males, then become dull grey blue all over. More commonly, the teneral female is bright orange and black, donning a grey blue cloak at maturity.

Flight Period: Mid-May through early September.

Range: Eastern: Newfoundland, Nova Scotia west to Manitoba and Minnesota, south to Texas and South Carolina. On Cape Cod, it is abundant everywhere.

Comments: I. *verticalis* is one of the first damselflies on the wing each spring and has one of the longest flight periods. It is the most common local species in its genus, quite abundant on the marshy borders of ponds, lakes and slow rivers. Adults perch in the shelter of shoreline vegetation or on floating plants. They fly slowly and are not difficult to capture. Pairs do not fly in tandem, so females are often seen on the ponds, ovipositing alone.

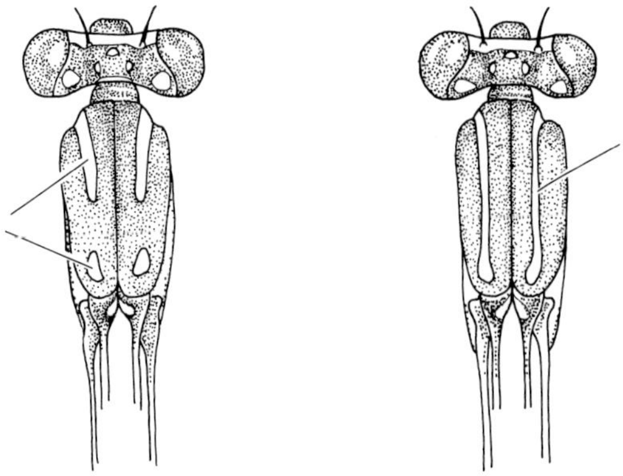

Figure 21. **Fragile Forktail** (*I. posita*), *left, and* **Common Forktail** (*I. verticalis*), *right, showing details of head and thorax.*

Green Damsels or Sprites (Genus *Nehalennia*)

These are delicate, very tiny damselflies with exquisite metallic green and turquoise color. There are 3 species in North America, all inhabitants of boggy waters. Of the 2 locally occurring species, neither is particularly common here.

SPHAGNUM SPRITE (*Nehalennia gracilis*) **Plate 1**

Description: A tiny, slender damselfly measuring 26-28 mm. (1 in.). The face is pale blue, the top of the head black with subtle metallic green lustre. There are no postocular spots. The thorax and slender abdomen are dark metallic green, except for the tip of the abdomen, which is pale blue. Females, though very much like the males in color, are noticeably stockier. The Sedge Sprite (N. *irene*) is very similar to N. *gracilis*, but the blue of abdominal segments 9 and 10 is interrupted by a pair of shiny black spots.
Flight Period: June and July.
Range: Maine and Nova Scotia west to Wisconsin, south only to Missouri and Ohio east to New Jersey and Massachusetts. On Cape Cod, both *Nehalennias* are rather scarce, with a few scattered populations here and there.
Comments: *Nehalennia gracilis*, small and slight of stature, nevertheless may become very abundant at sphagnum bogs. I found it numerous at a Brewster bog where it flew low over the carpet of wet *Sphagnum*, staying within the shelter of bog plants. Many pairs were also observed perched on the stems of sedges, particularly during late June and early July. N. *irene* is quite rare on the Cape, found occasionally on boggy edges of ponds.

The Dragonflies
(Suborder Anisoptera)

The Darners
(Family Aeshnidae)

These are some of the largest and most strikingly colored dragonflies. In general form they are of superb aerodynamic design: long slender abdomens and long, rather narrow wings. Darners are the super heavyweights of the dragonflies, many measuring over 3 inches in length. The goliath Hero Darner (*Epiaeschna heros*) is over 3½ inches long with a wingspan of nearly 5 inches.

The combination of robust size and torpedo-like flight makes the Darners a challenge to collect. But oh, what a temptation they present to the collector! Darners are high, swift fliers, often cruising along at tree-top level, and capable of reaching speeds of 30 miles per hour or more. A long-handled net with extra wide opening is a must for collecting these aerial acrobats. The Darners are powerful and voracious predators which consume tremendous quantities of insects. Their jaws are strong and sharp, and if mishandled they will give a painful bite. Adults may wander far from the breeding site in search of food or new habitat. At least one species, the Green Darner, (*Anax junius*), is migratory. Feeding swarms of Darners of mixed species are often seen at twilight in wild, spiralling flight amid thick clouds of mosquitos and gnats.

Adults oviposit in the stems of emergent or floating plants, such as rushes (*Juncus*) and pond lilies (*Nymphaea*). The nymphs are long, slender-bodied, and nicely patterned with cryptic colors which render them nearly invisible to their prey. They clamber about underwater plant stems, and will eat almost anything they can catch, including smaller members of their own group, fish and tadpoles.

Adult Darners are vividly colored, embellished with a kaleidoscope of intricate green, blue or yellow markings against dark brown or black ground color. Their eyes are enormous, larger than those of any other Odonate group. They wrap around the insect's head like a helmet, meeting on top for some distance. Identification of adults depends upon color pattern, wing venation and abdominal appendage shape. With a little study, the keen-eyed observer should have no trouble differentiating between the various genera. Of the 38 North American Darner species, 12 have been recorded on Cape Cod. It may be the giant Darners which instilled in humans their unwarranted fear of dragonflies and created the legends associated with them (see page 8). Though their size is frightening to some, the Darners are wonderfully handsome insects, completely harmless and extremely beneficial as predators of other insects, such as mosquitos.

Blue Darners (Genus *Aeshna*)

This is a large genus of robust, swift-flying dragonflies, many of which are autumnal and northern in distribution. Blue Darners are exquisitely colored insects, dark brown or black in background with a complex and diverse mosaic of cool blue green or azure spots and stripes to contrast. The spots of the abdomen and stripes of the thorax are distinct and precise, subtly different from species to species. In some Blue Darners, the base of the abdomen is constricted, giving the dragonfly a "wasp-waisted" appearance. Adults have a heavy black T-shaped mark on top of the head which varies in shape with species.

Determination of species can be difficult on the wing, but in hand the distinctively shaped paddle-like appendages of the male and rich blue and black mosaic coloration simplify the task.

I have spent countless hours pursuing these dragonflies, often merely for the thrill of the chase. *Aeshnas* are tempting but uncooperative targets: their large size and remarkable agility on the wing make them a challenge to net. Of the 25 species occurring in North America, I have recorded six on Cape Cod. These include the very common Spotted Blue Darner (*Aeshna clepsydra*) and the handsome Black-tipped Darner (A. *tuberculifera*). The lovely Spring Blue Darner (*Aeshna mutata*) is a little known Massachusetts Endangered Species which has been collected at just one Cape Cod site.

SPOTTED BLUE DARNER (*Aeshna clepsydra*) **Plate 3**

Description: *Aeshna clepsydra* is a small, slender, wasp-waisted species with a convoluted sky blue pattern on the sides of the thorax. Males measure from 66-70 mm. (2¾ in.), but local females are often greater than 70 mm. in length. The eyes and face are pale blue or bluish green and there is a heavy black cross stripe on the face. The sides of the thorax are marked with a complex, irregular pattern of pale blue to blue green spots and bands. This intricate mosaic is particularly distinct against the dark background color, and makes the species one of the easiest to recognize. The abdomen, swollen on segments 1 and 2, is sharply constricted on 3. The species name *clepsydra*, meaning "hourglass," is a direct reference to this trait. The abdomen shows an elaborate pattern of pale blue spots on a rich brown background. Male appendages are long, slender and gently rounded, the superiors paddle-like in appearance. Females are colored much like the males.

Flight Period: Early August through mid-September.

Range: Eastern: New England and Nova Scotia south to New York and Virginia, west to Indiana, Iowa, Wisconsin and Michigan. *Aeshna clepsydra* is our most abundant Blue Darner, known from all towns from Falmouth to Wellfleet, probably on to Provincetown.

Comments: *Aeshna clepsydra* is one of the easiest *Aeshnas* to capture. It shows a peculiar interest in the dense stands of Military Rush (*Juncus militaris*) encircling many of our Coastal Plain Ponds. I have my best luck standing in or near these emergent forests, where *Aeshna clepsydra* sidles casually along, wandering in and out among the tall stems of the rushes. It flies slowly compared to the bullet-like speed of its relatives, and habitually returns to favorite haunts where it will stop and hover.

BLACK-TIPPED DARNER (*Aeshna tuberculifera*) **Plate 3**

Description: A big, strapping blue *Aeshna*, much larger than the previous species at 76 to 82 mm. (3-3¼ in.). The face is light bluish green, the thorax finely marked with precise, complete stripes. The lateral stripes, azure blue towards the top grading into green below, are simple, parallel-sided and regular in shape. They are beautifully defined against the reddish brown ground color. The legs are red brown at the base, becoming black on the last two segments. Wings show no color except the red brown stigmas and a slight tinge of amber in some individuals. This is another wasp-waisted species with swollen basal abdominal segments and constricted middle segments. The abdomen is elegantly marked with a double row of large blue spots dorsally and a second row of small spots very low on the side. Segment 10 is black and the appendages are paddle-shaped in profile.

Flight period: Late July through early September.

Range: Northeastern: Maine, Nova Scotia, Quebec and Ontario west to Wisconsin and Washington, south to Pennsylvania and New York, Connecticut and Virginia. *Aeshna tuberculifera* is known locally from Falmouth to Wellfleet, and very likely occurs north to Provincetown.

Comments: This species frequents the grassy shores of ponds, much like the previous species. It too repeatedly returns to cruise over the emergent grasses and rushes, but is somewhat more wary than the Spotted Blue Darner and flies with greater speed. On a number of occasions I have missed *Aeshna tuberculifera* on the first swing of the net, but have been successful with a quick back-sweep as the insect seemingly returns for another look! This species also has an intriguing habit of disappearing "into thin air" as it rockets skyward or drops like a rock into a jungle of waist-high rushes.

SHADOW DARNER (*Aeshna umbrosa*) — Plate 3

Description: This is a slender, dark species, the male with distinctively shaped abdominal appendages. Adults measure 68-75 mm. (3 in.) in length. The face is pale green in most, without a black cross stripe. The base color of this Darner is deep reddish brown, highlighted by spots and stripes of green and blue. The usual 2 lateral bands on the thorax are very narrow, greenish yellow or yellow, bordered by black. The abdomen is spotted with blue or greenish blue as usual in this genus, but the spots are quite small, making the insect appear more dark overall. Another unusual feature of this species is the paired large blue spots on the underside of abdominal segments 4 through 7. The abdomen is very long and slender, with a strong constriction of segment 3. Each superior appendage in the male bears a single spine just before the tip of the "paddle", which points downward and rearward (see Figure 11).

Flight Period: *Aeshna umbrosa* flies from mid-summer into fall, first visible here in July and not uncommonly seen through the second week of October.

Range: Transcontinental in Canada, but generally eastern in the U.S.: Maine south to Georgia, west through Mississippi and Oklahoma, north to Minnesota. I have recorded this species in Dennis, Harwich, Brewster and Truro, but it probably occurs Cape-wide.

Comments: *Aeshna umbrosa* is very widespread and abundant throughout its North American range, where it haunts streams, pools and ditches. I frequently encounter it along shady woodland trails and old bog roads. This slender, dark Aeshnid is quite distinctive in behavior, often flying late into the afternoon and evening hours. Oddly enough, many of the specimens I have collected have come from inside buildings, and at least twice we have had males enter the Museum. One individual flew into the second grade classroom at the Truro Elementary School, perched hanging downward from the ceiling and remained there for a day or two, where it eventually died. This species is one of the very last dragonflies on the wing, often flying well into October.

Green Darners (Genus *Anax*)

The Green Darners are distinguished from other darners by their uniform, vivid green thorax which lacks the spotting or striping typical of other genera. Of the four North American species, two occur on Cape Cod: one widespread and very common, and the other more rare and local.

COMMON GREEN DARNER (*Anax junius*) — Plate 4

Description: Adult Green Darners measure 70-80 mm. (3-3¼ in.) from head to tip of abdomen, the females slightly larger and chunkier than males. The eyes and face are greenish yellow, and the top of the head has a distinctive black bulls-eye pattern (a black spot surrounded first by a yellow ring and then blue). Though not a practical field mark, this spot will separate *Anax junius* from closely related species. The thorax is bright green, the abdomen deep azure blue (the color of a fall sky) and black. The blue predominates at the base, giving way to black towards the tip. Wings are clear, sometimes washed with pale amber. Females are similar to males, but the abdomen is stocky and brick red in color.

Flight period: June through September.

Range: *Anax junius* is the most widely distributed Darner of its genus. It is known from almost every state including Alaska and Hawaii, and is common in Canada, the Caribbean, Mexico and even Asia. On Cape Cod, it ranges from Woods Hole and Sagamore to Provincetown.

Comments: The Green Darner is the most abundant species in its genus, and on Cape Cod it ranks as the most abundant of all Darners. It is a very familiar summer dragonfly, frequently encountered around the shores of shallow, heavily vegetated ponds. Though its large size makes it seemingly immune to predators, both adults and nymphs are regularly taken by other insects including their own species, fish, frogs, turtles and birds. I once observed a beautiful male flying low among emergent plants, apparently oblivious to possible danger. As he meandered lazily along, I thrilled to the brilliance of his colors, the setting sun and the peace of the evening pond. Without warning and much to my dismay, a 10-inch large-mouthed bass leaped clear of the water and snatched the Green Darner from the air, falling back into the pond, dinner secured, with a resounding splash.

The Common Green Darner is a powerful, active insect, spending virtually all of its daylight time on the wing. Feeding swarms of Green Darners are frequently encountered along coastal dunes, in fields, bogs and backyards where they can effectively control pest insects.

Females oviposit in water lilies and emergent plants, with the male in tandem. The nymphs can develop to maturity in less than 6 months if food is abundant.

Adults of this species are wide-ranging in their activities, and are known to migrate along the Atlantic Coast in huge groups, one generation drifting northward with the birds in the spring and their eventual offspring spinning southward the following autumn.

LONG-LEGGED GREEN DARNER (*Anax longipes*) — Plate 4

Description: This is easily our most spectacular dragonfly: a magnificent, strikingly colored red giant, solid and robust in stature. The Long-legged Green Darner is nearly the size of a small bird (75-87 mm. or 3-3½ in.) and noticeably larger than *Anax junius*. Males, with blazing scarlet abdomen and grass green thorax, are both unmistakable and unforgettable even at some distance. Their legs are extremely long (hence the Latin "*longpipes*"), black at the base, brick red thereafter. The wings are hyaline, though like *Anax junius* they may be tinged with amber. The greenish yellow head has no bulls-eye mark like that of A. *junius*. Female Long-legged Green Darners are chunkier of body and brick red in color, similar to the hues of the female A. *junius*.

Flight Period: June through September on Cape Cod, with peak breeding from mid-June to mid-July.

Range: Primarily Atlantic Coastal Plain: Massachusetts south along the coast to Florida and Texas, inland to Ohio. In Massachusetts, it is most numerous on Cape Cod, where it ranges from Falmouth to Brewster. Its absence from outer Cape ponds is puzzling, since suitable habitat does exist there. *Anax longipes*, at the northern fringe of its range, is quite uncommon in Massachusetts, where it is a Species of Special Concern.

Comments: The Long-legged Green Darner is a fleet, powerful insect which flies with an easy, fluid grace. Its bird-like size, steady and purposeful flight, and resplendent colors make it the lord of the Coastal Plain Pond. It frequents shallow, sandy ponds with ample shoreline vegetation in the form of rushes (*Juncus* sp.), Plymouth Gentian (*Sabatia kennedyana*) and Pickerelweed (*Pontederia cordata*). Long-legged Green Darners often cruise over these near shore emergent jungles, but range widely over the pond. Females oviposit in the stems of water lilies or other aquatic vegetation, and are unaccompanied by the male.

This is one of the strongest dragonfly species, and it is rarely seen at rest. The swiftness of its flight accounts in part for its rarity in collections. In fact, in all my years of studying dragonflies I have managed to catch only four specimens. Long-legged Green Darners prey on small insects, but are often seen heading for the tree-tops with larger prey items. One gargantuan male I watched at a Hyannis pond appeared to have a small bird locked in its grasp. I had to look twice to convince myself that its prey was most certainly not a sparrow, but rather three other dragonflies, each approximately half the darner's size! Indeed, the Long-legged Green Darner rules the airspace over its Coastal Plain Pond habitat, but often runs a gauntlet of aggressive dragonflies in its daily activities. Interactions with other species result in remarkable high-speed chases over the pond. Common Green Darners (*Anax junius*), Golden-wings (*Libellula auripennis*) and Damson Skimmers (L. *incesta*) regularly attempt to drive off the Long-legged Green Darner, but are successful only temporarily.

Other Darners

Four additional Darner genera are represented on Cape Cod, with one species in each genus. All are relatively rare locally, and thus have not been described individually in this book. Two species with stream and river affinities are very rare here. The Stream Darner (*Boyeria vinosa*), perhaps the most drably colored of the Darners, is uniformly brown with 2 round yellow spots on the sides of the thorax and less conspicuous spotting on the abdomen. The Springtime Darner (*Basiaeschna janata*) is also brown, but spotted with lovely pale blue and green. One sphagnum bog species, the smartly colored Harlequin Bog Darner (*Gomphaeschna furcillata*), is also recorded from scattered Cape locations. The colossal Hero Darner (*Epiaeschna heros*), the monarch of all darners, measures to 94 mm. (3¾ in.). It is a glorious beast with ocean blue eyes, bright green stripes on the brown thorax and very fine green rings on the abdomen. I have caught only one individual of this species on Cape Cod. It is not common here, though its preferred habitat of small ponds and swamps is plentiful. The single spectacular male taken in Wellfleet was part of a feeding swarm of Common Green Darners (*Anax junius*) which were gorging themselves on deer flies, green-heads and bees. It had bits and pieces of various meals stuck on its thorax, a fly wing on its side and a grasshopper leg underneath.

Clubtails

(Family Gomphidae)

Clubtails generally frequent the clear, oxygen-rich waters of streams and rivers. The family is a giant one, with almost 100 North American species, but only three common representatives on Cape Cod. Because of the lack of swift-running streams and rivers here, our Clubtail fauna is limited. However, many species, including those described here also inhabit the sandy shores of our Coastal Plain Ponds and larger lakes and thus may become quite widespread where this habitat is available.

This is a primitive group of dragonflies whose members show their ancient ancestry in the wide separation of the eyes and various features of wing venation. Clubtails are so-named because of the distinct flaring of the tip of the abdomen, a trait most conspicuous in males. The degree of flaring varies between species, but the presence of the abdominal club is a reliable feature of the family. Behaviorally, the Clubtails are also distinctive. Adults usually perch on the ground or on rocks and their legs are short and sprawling for this purpose. When they fly, they generally stay close to the water, cruising up and down the river or lake edges. A camouflage pattern of yellow, green and brown is surprisingly effective in most species, but if they can be spotted while at rest on the ground, adults are not difficult to capture. The contrast of dark body color in many species against pale sand or rock is fairly easy to spot.

Closely related species are identified by the color pattern of the thorax and abdomen, the shape of the abdominal appendages and, to a certain degree, by the size of the abdominal club. Adults do not fly in tandem and females are on their own during oviposition. As a result, they are frequently harassed by males and have become very secretive in habit. Nymphs are oddly shaped creatures, adept at burrowing in stream or lake bottom sediments. Members of this family, requiring oxygen-rich waters, are fairly intolerant of pollution.

Spinylegs (Genus *Dromogomphus*)

BLACK-SHOULDERED SPINYLEG (*Dromogomphus spinosus*) **Figure 22**

Description: A slender Clubtail with small head and very long, spiny hind legs, this is the largest representative of the family recorded on Cape Cod. Males measure 57-59 mm. (about 2¼ in.). Females are larger (61+ mm.) and chunkier with only slight flaring of the abdomen. The face and head are yellow, the eyes beautiful blue green. Two narrow stripes on the front of the thorax form a loose triangle with the yellow collar. The sides of the thorax are yellow, sometimes separated into two stripes, otherwise confluent. The exceedingly long hind femora are armed with sharp elongate spines (hence the specific name *spinosus*). A narrow yellow stripe runs lengthwise along the shining black abdomen, ending as two triangular spots on segments 8 and 9 and a squarish spot on 10.

Flight Period: July and August

Range: Southeastern Canada west to Wisconsin and Texas, south to Florida. *Dromogomphus spinosus* occurs throughout New England, but is known only from Falmouth to Brewster on Cape Cod.

Comments: This fine Clubtail is relatively uncommon in our area, having been collected at only 5 locations on Cape Cod. Its preferred habitat, like others of the family, is clean rivers and streams. Here, it is found on large, clear-watered lakes with sandy substrate. The Spinyleg is a secretive dragonfly which is often encountered flying about in woodlands some distance from the breeding pond. Hence, it is probably more common here than records indicate. Adults perch high in trees, swooping down to patrol broad areas of the pond. Individuals fly repeatedly over the same region, returning to the tree-tops periodically to rest.

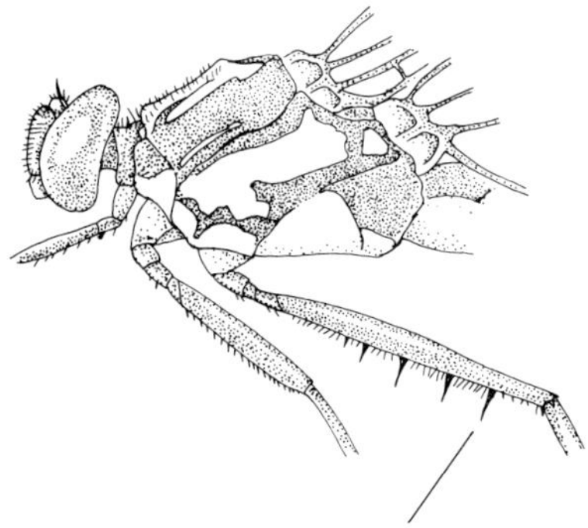

Figure 22. **Black-shouldered Spinyleg** (D. *spinosus*). *Detail of head, thorax and legs.*

Common Clubtails (Genus *Gomphus*)

This enormous, complex genus includes about 36 North American species, of which only 4 are recorded from the Cape. Of these four, two are known only as historical records: one from the late 1800s, another from the 1920s. The following species is our only common *Gomphus*, though there is a recent record of the rare Brotherly Clubtail, G. *fraternus*.

LANCET CLUBTAIL (*Gomphus exilis*) **Figure 23, Plate 4**

Description: The least striking of our Gomphid species, *Gomphus exilis* is a small, dainty, brownish Clubtail with yellowish green markings. Most Barnstable County specimens measure from 38-42 mm. (1½-1¾ in.). The face and top of the head are pale yellowish brown. The thorax shows the usual Clubtail striping, with 2 narrow dorsal pale stripes converging towards the base of the wings and 2 broader pale stripes visible on the sides. The legs are brown with yellow on the outer side of the tibia and tarsi. This Clubtail is very slender of abdomen and shows only a gentle flaring of the end segments. It is dark brown with an elongate lancet-shaped yellow mark covering the length of each segment. The sides of the abdomen are marked with a variable pattern of pale spots. Females are similar to males, but do not have a distinct club.

Flight Period: Spring to mid-summer: early May to early August.

Range: Eastern, from Manitoba east to Nova Scotia, south to Florida and west to Texas and Wisconsin. On Cape Cod, Falmouth and Bourne through all towns to the Cape tip.

Comments: This is our most common Clubtail, found in abundance at sandy bottomed ponds and freshwater rivers and brooks. Like others of its family, *Gomphus exilis* is most often seen resting on the ground, on logs or on rocks. It may wander after emergence, and I have found pairs at considerable distances from the breeding habitat.

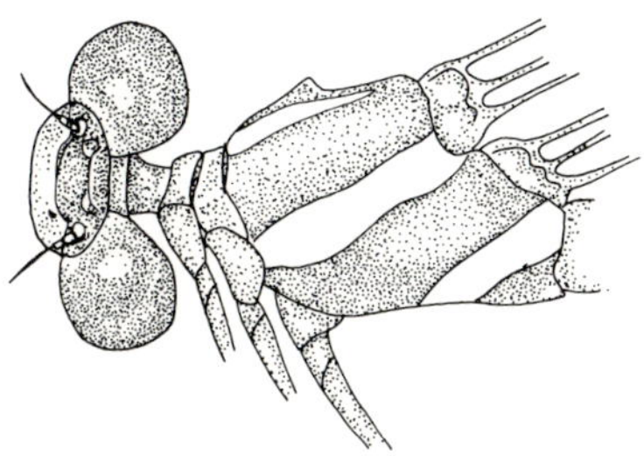

Figure 23. **Lancet Clubtail** (G. *exilis*). *Detail of head and thorax.*

Sand Dragons (Genus *Progomphus*)

SAND DRAGON (*Progomphus obscurus*) **Plate 4**

Description: A handsomely marked, medium-sized Clubtail at 50-53 mm. (about 2 in.). *Progomphus obscurus* has short stocky legs and a relatively long slender abdomen. Its dark chocolate brown (almost black) body is nicely marked with contrasting sunny yellow spots and stripes. The face is pale with greenish yellow eyes which become brown in dried specimens. The thorax shows two narrow yellow stripes on the front and a more complex sinuous pattern on the sides. The short legs are dark brown with a rectangular yellow spot on the outer side of the femora. The wings are clear with long narrow brown stigmas and a very small brown wash at the base. The thickened costa vein on the leading edge of the wing is yellow. The abdomen, richly brown in background, is gaily marked with large triangular yellow spots on top of segments 2 through 7 and a rounded lateral spot on 7. Spotting is reduced to very small pale lateral marks on the gently flared end segments. The superior appendages are bright yellow, the inferiors black. Females are colored much like the males, but lack the abdominal club.

Flight period: Early to mid-June through early August.

Range: New Hampshire and Massachusetts west to Colorado and Wyoming, south to Texas and east to Florida. On Cape Cod, Falmouth to Truro. The species may not occur in Provincetown, due to the lack of appropriate habitat there.

Comments: This is one of our most common Clubtails, found at almost any clear, sandy-bottomed lake or pond and sometimes fresh streams or rivers. While it prefers rivers and streams in much of its range, the Sand Dragon finds the loose sediments of numerous Coastal Plain Ponds equally suitable and certainly more available on Cape Cod. Nymphs tunnel along just beneath the sand, leaving obvious meandering trails like miniature mole tunnels. These trails, thick along the shores of larger ponds, are a reliable indicator of the presence of this species. Adult Sand Dragons light on the ground on sunny exposed beaches, with abdomen raised slightly. They are surprisingly wary of human approach, flushing quickly and sailing smoothly over the water within a few inches of its surface. Mating lasts more than 10 minutes, but pairs are rarely seen on the ponds since they usually rest in trees or shrubbery. Females oviposit along the shores of ponds or in relatively quiet stream riffles, by dunking the abdomen in water and releasing individual eggs. This is apparently the only species of Clubtail where the male guards his ovipositing mate.

Emeralds
(Family Corduliidae)

These are medium-sized, dark and elegant dragonflies, slender and gracefully built, many with sparkling emerald eyes and shining bronzy-green bodies. Members of this family, though closely allied to the giant Libellulidae family, are distinctly different in appearance. Emeralds are slender-bodied, dark and metallic in color and possess characteristic wing venation. The end of the anal loop of the hind wing is not noticeably foot-shaped, lacking development of a "toe" or "heel" portion as in the Libellulidae (Figure 9).

Emeralds can be found in a variety of aquatic habitats, from streams to mucky pools, ponds, sphagnum bogs and sandy lake shores. Nymphs sprawl in bottom debris or clamber about in aquatic vegetation. On Cape Cod, I have recorded only 5 rather common species, though the potential for others does exist. The family itself is quite large, with nearly 50 North American species. Many of its members are of northerly distribution or inhabit specialized habitats not present on Cape Cod.

Little Emeralds (Genus *Dorocordulia*)

These are small, dark dragonflies similar in general appearance to the *Somatochloras* (p. 55). The genus contains only 2 eastern species, one of which is quite widespread on Cape Cod.

PETITE EMERALD (*Dorocordulia lepida*) **Figure 24**

Description: A charming, dainty little dragonfly, the smallest of our Cape Cod Corduliids at 37—41 mm. (1⅝ in.). Colored much like the *Somatochloras*, the Petite Emerald has brilliant green eyes and face, hairy brown thorax reflecting metallic green lights and slender black abdomen spotted with yellow on the basal segments. The abdomen, rather heavy at segment 2, becomes slender through the middle segments, flaring again at its tip. The wings are clear with black stigmas. Females resemble males but the abdomen is much heavier.

Flight Period: June through August.

Range: Tightly northeastern: Nova Scotia and New England south to New York, New Jersey and Maryland; widespread on Cape Cod, where it occurs from Woods Hole and Falmouth to the Cape tip.

Comments: The Petite Emerald is quite common on Cape Cod, inhabiting boggy ponds, cranberry bogs and swamps, and our few sphagnum bogs. I have seen unattended females ovipositing in the irrigation ditches of working cranberry bogs and have had water-logged specimens brought to me from backyard swimming pools. Males are active and territorial, their unmistakable emerald eyes flashing as they hover over a warming summer pond.

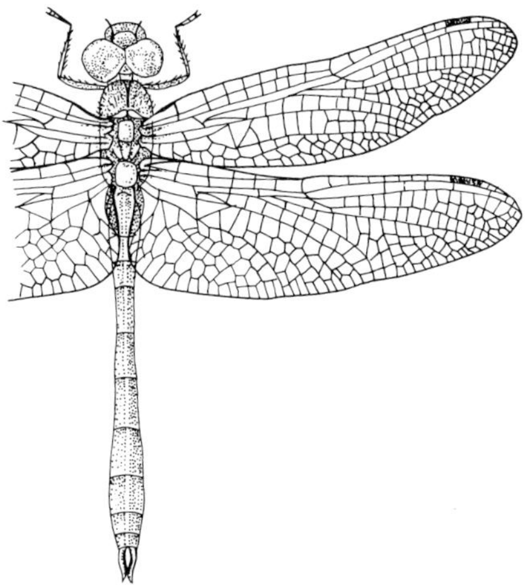

Figure 24. **Petite Emerald** (D. *lepida*)

Baskettails (Genus *Epitheca*)

Baskettails are medium to large dragonflies, dull brown in color and quite hairy on the thorax. Many species are very similar in appearance and must be separated by the shape of the abdominal appendages. Fortunately we have only 2 common, easily identified species on Cape Cod. The English name Baskettail relates to the divided plate beneath segment 9 of the female's abdomen which allows her to carry an entire batch of eggs at once. The ball of eggs is dropped into the water with a quick dunking of the abdomen, and thus the female is exposed to predation during oviposition just once. The generic status of this difficult group remains highly debatable: entomologists place its members in the genus *Epitheca*, *Tetragoneuria* or *Epicordulia*.

COMMON BASKETTAIL (*Epitheca cynosura*) **Plate 5**

Description: A medium-sized dragonfly (37—41 mm. or 1⅝ in.) with pearly blue eyes, brown thorax covered with long grey hairs and brownish black abdomen dappled with pale gold lateral spots. The wing pattern is extremely variable: spotting is limited to the basal third of the hind wings, but in some individuals a heavy brown triangular spot reaches along the leading edge of the wing nearly to the nodus. In others the spot is more diffuse and transparent in places or it may be irregular with heaviest color confined to the wing veins. In a few individuals the brown is so reduced that it does not reach the triangles of the hindwing.

Flight Period: Late May through July, most abundant on Cape Cod in June.

Range: Maine and Nova Scotia, southern Quebec and Ontario west to Minnesota, south to Texas and Florida. Recorded in every Cape Cod town.

Comments: If you have a medium-sized brown dragonfly hawking mosquitos in your yard in June, chances are it is a Common Baskettail. This is one of the most abundant Odonates in eastern North America and certainly Cape Cod's most abundant spring dragonfly. It prefers relatively still waters including lakes, muddy ponds, swamps, marshes and lazy streams. Most of the population emerges within a few days time, as is typical of many explosive spring species. At one tiny Brewster pond I counted over 100 exuviae in an area of only a few square yards. Many plant stems had five or six empty skins lined up nose to tail, testimony to the enormous population there. Baskettails gather in large feeding groups along sunlit woodland roads and clearings and even along well-traveled highways. While swarm-feeding, their flight is straight-lined and purposeful, with quick directional changes and little or no hovering. Adults perch vertically on low vegetation, soaking up the warming June sun. Males are territorial on the breeding habitat, nervously patrolling, hovering and maneuvering.

THE PRINCE (*Epitheca princeps*) **Figure 25**

Description: The Prince is unmistakable for its large size (63—68 mm. or 2½-2¾ in.), regal green eyes and the unusual brown spotting of its wings. Each wing shows an elongate spot at its base, a more irregular spot at the nodus, and a third at the tip. The brown thorax is thickly clothed with hair, concealing in part a faint bronzy iridescence. The long, slender abdomen is nearly black in color and marked with paler brown along its sides. Females closely resemble the males.

Flight Period: Late June through August.

Range: New England, southern Quebec and Ontario, west to Minnesota, south to Texas and Florida. The species is widespread and not uncommon on Cape Cod, where it is recorded from the Canal to Truro.

Comments: This large, Darner-sized Corduliid inhabits clear sandy-bottomed lakes and quiet rivers. Females drop blobs of eggs into clean open waters, unaccompanied by a male. The Prince behaves much like the Darners: males patrol well-defined territories and if they wander from these regions they will return a short time later to take up the "beat" once again. Were it not for the fact that these patrols are usually just far enough off-shore to be unreachable on foot, the Prince would be an easy catch. I have better luck netting adults away from the ponds where they gather in swarms of 5 or 6 individuals to forage on small insects. Powerline cuts, roads and fields are common feeding areas. Unfortunately, due to its habit of feeding along open roads, the Prince is often "road-killed" by automobiles. Its flight is generally lofty with periods of flapping flight interrupted by short glides on still wings. I once watched a big, husky female flying over a road well above reach, refusing to drop down so I could take a closer look. When a male came along, forcing her down to the pavement, I plopped my net over both of them! The Prince often forages at twilight, a ghostly brown dragon aloft on spotted wings. Nearly invisible in the fading light, it makes short runs in one direction, coming quickly about and working a remarkable labyrinthine flight pattern that is very difficult for the human eye to follow.

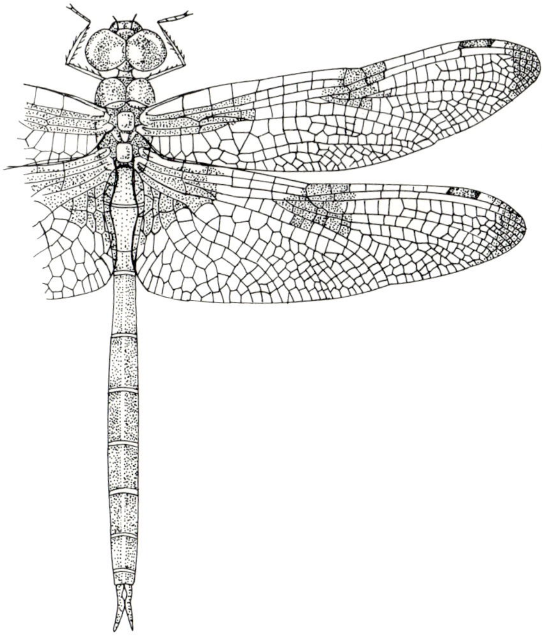

Figure 25. **The Prince** (E. *princeps*)

Striped Emeralds (Genus *Somatochlora*)

This is a large group of Emeralds with 26 North American species, predominantly northern in distribution. All are slender-bodied with dark, metallic color and various spots or stripes on the body. I have recorded only 2 species on Cape Cod, the most common of which I will describe here.

CLAMP-TIPPED EMERALD (*Somatochlora tenebrosa*) **Figure 26**

Description: A large, darkly elegant dragonfly, the male with very distinctive abdominal appendages. Adults measure 54—60 mm. (2¼ in.) and females are a big larger than males. In life, the eyes are richly emerald, fading to brown in death. The brown thorax has two sunny yellow lateral bands and shimmers with a greenish lustre. The abdomen is long, slender and black, with yellow spots on the sides of segment 2 and 3. The peculiar clamp-like appendages of the male consist of sharply arched superiors and smoothly upcurved inferiors. The vulvar plate of the female, large and triangular in shape, points downward at a deep angle from segment 9.

Flight Period: July through September.

Range: Nova Scotia, New England to Michigan, south to Florida, west to Oklahoma. I have taken S. *tenebrosa* only in Dennisport and Brewster, but have seen it in a few other locales on Cape Cod. It is quite common off-Cape.

Comments: These are high, strong fliers which can be exceedingly difficult to capture. They forage along shaded roads through Atlantic White Cedar Swamps or in openings in the forest canopy, where their dark body colors are lost against the deep evergreen foliage. Adults perch hanging downward from branches. They are most often associated with Atlantic White Cedar Swamps and streams, where they fly low, regular beats during breeding. I have several times observed a peculiar behavior from this Emerald in a Dennisport wetland, where a little-used road runs through an Atlantic White Cedar Swamp. The road floods consistently in one area after heavy rains. Males, flying along the road or flushing from a perch, dove directly into the pool of water, sometimes struggling to become airborn after a dunking. The reason for this behavior is unclear, although several possibilities exist: the dragonfly may be attempting to pick prey items off the water (examination revealed little or no insect activity on the puddle's surface); it may see its own reflection and attempt to drive the "intruder" away; or perhaps S. *tenebrosa* is merely cooling off on a hot August day! Curiously, I have observed other insects, such as wasps, performing the same strange water-dive.

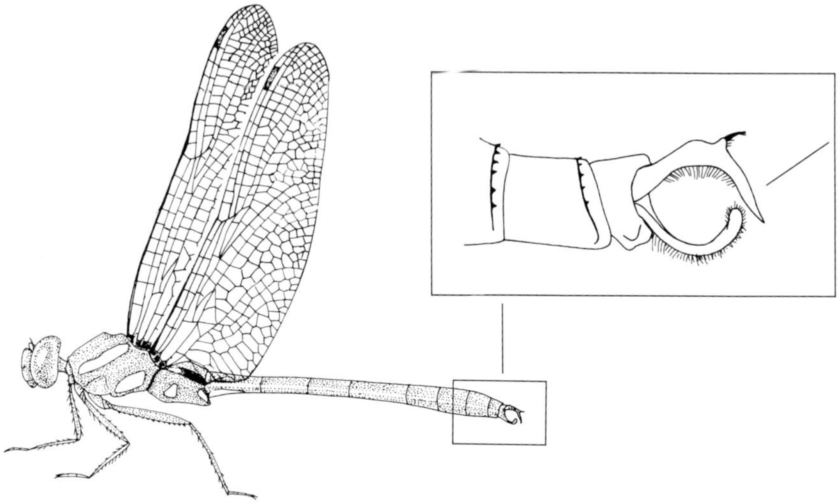

Figure 26. *Left*: **Clamp-tipped Emerald** (S. *tenebrosa*).
Right: *Detail of clamp-like appendages on male abdomen.*

Cruisers
(Family Macromiidae)

These are robust, dark-colored dragonflies with a single belt of yellow encircling the thorax and spots of yellow on the abdomen. They are swift and powerful on the wing, capable of sustained long-distance flight. Adults cruise along rivers or the edges of ponds and lakes, and can be exceedingly difficult to net. The nymphs are bizarre in appearance, designed for sprawling undetected on sandy river or lake bottoms. A camouflage of dappled browns, wide flattened body, long spidery legs and prominent horn between the eyes are characteristics of Macromiidae larvae. The family contains just two genera with 10 North American species. Only 2 occur on Cape Cod.

Brown Cruisers (Genus *Didymops*)

This small genus, with two North American species, is represented here by the more common of the two, *Didymops transversa*. Brown Cruisers inhabit sandy windblown lakes and ponds or rivers. Females oviposit by dunking the abdomen repeatedly in quiet shallow water.

STREAM CRUISER (*Didymops transversa*) **Plate 5**

Description: A fine chestnut brown dragonfly with a cross stripe of yellow on the face, a single yellow band wrapping completely around its rather hairy thorax, and yellow spots on the top of abdominal segments 2-8. These are medium-sized, stocky dragonflies, measuring 52—57 mm. (2¼ in.). The wings are clear except for a very small brown basal mark, and the legs are unusually long. The sunny yellow spots on abdominal segments 2—6 are divided by a narrow brown streak, making them appear as two. The tip of the abdomen is widened into a club somewhat reminiscent of the Family Gomphidae, and the superior appendages of the male are bright yellow.

Flight Period: Late May through early August.

Range: Nova Scotia, Ontario and Quebec to Minnesota, south to Texas and Florida. The Stream Cruiser is widespread and common on Cape Cod, where it frequents windswept lakes from the Upper Cape to Truro.

Comments: Adults of this species fly very low over the water's surface, with the clubbed abdomen slightly raised. I often come upon them, both teneral and mature, flying about in the woodlands near ponds or lakes. Here too, they fly low, within a few feet of the ground, sailing easily through the thick underbrush. They are extremely well camouflaged and usually perch just above the ground, where they are nearly invisible against the underbrush and brown leaf litter. I recall a particularly frustrating hour of pursuit during which I flushed *Didymops* several times, chased him through tangles of catbriar and blueberry, only to have him drop into the brush, invisible once again. Irritated but still curious, I looked down and there he was hanging from a lowbush blueberry, inches from the ground and inches from my feet!

River Cruisers (Genus *Macromia*)

These are big, dark river monsters with bold black and yellow bodies and shimmering emerald green eyes. Of the 8 North American species, only one occurs on Cape Cod and it is really quite rare here.

SWIFT RIVER CRUISER (*Macromia illinoiensis*) **Plate 5**

Description: This is a dark, stunningly handsome insect with exquisite green eyes and spots of primrose yellow on the abdomen. Local specimens range from 64—67 mm. (2½—2¾ in.): those of northern and western regions of the state are larger. The thorax, with subtle metallic green luminescence, is dark brown with a single bright yellow side stripe and a coat of fine pale hairs. The wings are hyaline with dark venation and stigma. The jet black abdomen bears vibrant yellow spots on the top and sides of segment 2, small paired dorsal spots on 3 and 4, a larger dorsal spot on 7 and a small narrow spot on 8. All other segments and appendages are wholly black. The abdomen is swollen at its base, slender through the middle segments and gracefully widened near the tip. Females are larger and chunkier, with more extensive yellow spotting on the abdomen.

Flight Period: Late June through the first few days of September.

Range: Nova Scotia to Manitoba south to Mississippi and Georgia, west to Kansas. It is known from Falmouth, Harwich, Brewster and Truro.

Comments: These are remarkably strong-flying dragonflies, with potentially vast home ranges. They are associated with rapidly flowing rivers, but more frequently with large sandy lakes in our area. I usually encounter adults as they rocket along western Massachusetts river corridors, which they use as aerial freeways. On Cape Cod, River Cruisers use woodland roadways as travel corridors. Most of my specimens have been collected along these "highways" where adults may travel long distances uninterrupted. Adults, though famous aerial wizards, are not always difficult to capture, particularly along rivers or roadways. Their flight is fast, straight-lined and unfaltering and they do not stop to hover or change directions frequently. Ovipositing females fly low and fast, occasionally tapping their abdomens on the surface to release eggs.

Common Skimmers
(Family Libellulidae)

More than half of Cape Cod's Anisoptera (dragonflies) belong to this enormous family. It is the largest of all Odonate groups, with 96 North American species, and ranges throughout the world. Members of the Libellulidae are common about quiet waters, many flourishing in eutrophic conditions. On Cape Cod, they are abundant in virtually any habitat type including saltmarshes, but are most numerous in muddy freshwater ponds, pools, marshes and other stagnant waters.

Adult Common Skimmers are flashy and colorful, but lack the metallic shimmer of the Corduliidae. The wings of many species are dappled with spots or bands of color. The wing venation is characteristic, the most notable feature being the distinctly foot-shaped anal loop of the hind wings (Figure 9). Common Skimmers have large eyes which, as in most dragonflies, meet on top of the head.

These are perching dragonflies which frequently alight on twigs or emergent plants. A few species, the Corporals and Whitetails for instance, also perch on the ground. Common Skimmers are so-named for their habit of skimming along the water's surface or just above shoreline weeds. Many species are ace fliers, capable of short bursts of great speed and astonishing maneuverability. Females drop eggs freely into pond waters, either alone or in tandem with a mate. Libellulidae larvae are difficult to separate from those of the Corduliidae, to which they are closely related. They are active, rapidly growing nymphs which thrive in the warm stagnant waters of Cape Cod ponds.

Spot-winged Skimmers or Pennants (Genus *Celithemis*)

Four of the 8 species in this genus occur on Cape Cod. They are small to medium-sized, slender in stature and most have ornately spotted wings. All locally occurring species fly during summer months at Coastal Plain Ponds with emergent rushes, grasses and other plants. Pairs oviposit in tandem, skimming low over open pond waters. The female taps her abdomen rhythmically on the surface, washing blobs of eggs into the water.

ELISA SKIMMER or CALICO PENNANT (*Celithemis elisa*) **Plate 6**

Description: A fine red dragonfly with rather short abdomen and long, spotted wings. Adults are small in size at 29—32 mm. (1¼ in.). In mature males the face is red, the eyes chestnut, the thorax reddish with black markings and a coat of velvety brown hair. The hind wings are dappled with a large brown and amber mark at the base, another smaller spot just beyond the nodus and a third at the apex. Forewing spots include a small brown spot midway between the base and nodus, another pale spot just beyond the nodus and a third at the wingtip. The stigmas are bright red. The jet black abdomen is spotted with a rich cherry red on top. Females, similar in pattern, are yellow where the males are red.

Flight Period: June through August.

Range: Southern Ontario, Quebec, south through the U.S. to Florida and west to Minnesota and the Plains states. The Elisa Skimmer is abundant and widespread on Cape Cod, ranging from Bourne and Falmouth to Provincetown.

Comments: This is the most abundant species of its genus on Cape Cod. I have found this gaily colored Skimmer in a wide variety of habitats here: from relatively clear, sandy lakes to very mucky shallow ponds, the slow parts of streams and even brackish marshes. Adults prefer weedy ponds but are adaptable to a great variety of situations.

HALLOWEEN PENNANT (*Celithemis eponina*) **Plate 6**

Description: The largest of our *Celithemis* species at 34—39 mm. (1½ in), C. *eponina* is ornately patterned in the traditional Halloween colors of orange and black. The long, broad harlequin wings are this species' trademark: the entire wing membrane is pale ochre, banded and spotted with dark ruddy brown, and the stigmas are crimson. The abdomen is black with orange yellow spots on top of segments 3 to 7. Mature males may appear almost red, while females are pale yellow.

Flight Period: July and August.

Range: Maine west through southern Ontario to Minnesota and Nebraska, south to Texas and Florida. *Celithemis eponina* occurs from Falmouth to Provincetown.

Comments: This striking, vividly colored Skimmer exhibits a distinctive fluttery, butterfly-like flight. It perches most often near the tops of emergent rushes and is active very early in the morning. It is not as common here as other Pennants, but is nevertheless a conspicuous fellow wherever it occurs.

MARTHA'S PENNANT (*Celithemis martha*) **Plate 6**

Description: A small, dark-colored dragonfly measuring 29—32 mm. (1¼ in.). Wing color is confined to a round, blackish basal spot on the hind wing, extending to the outer angle of the triangle and about ¾ the way from leading to trailing edge. In females and young males, the spot is gold and black. Younger adults are prettily marked with yellow on the sides of the thorax and bright yellow triangular spots on top of the black abdomen. At full maturity, males become ebony black from head to tail.

Flight Period: Late June through August.

Range: Northeastern: Nova Scotia, Maine, south along the coast to Virginia; also recorded from Pennsylvania and Ohio. On Cape Cod, it is known from Falmouth to Truro.

Comments: This small, dark phantom becomes especially numerous at shallow, productive Coastal Plain Ponds. Peak breeding occurs during the last part of July and the first part of August, when dozens of tandem ovipositing pairs can be seen bobbing along at the water's surface.

BANDED PENNANT (*Celithemis monomelaena*) **Plate 6**

Description: Much like Martha's Pennant in color, except for its many-spotted wings. This species is also slightly larger, to about 35 mm. (1⅜ in.). Immature males and females are dark brown and yellow with stripes on the sides of the thorax. Males become wholly black at maturity. The unusual variegated wing pattern appears as follows: a long narrow double brown stripe runs from wing base to nodus, followed by an irregularly shaped spot halfway between nodus and wing-tip, and a third covering the wing tip. Some males show an additional round basal spot halfway between leading and trailing edges of the hind-wing, while in others it is greatly reduced or absent. Females are similarly spotted, but the wing-tip spot is absent in most or reduced to a band below the wing-tip.

Flight Period: July and August.

Range: *Celithemis monomelaena* is similar to the southern C. *fasciata*, and indeed considered one and the same by some Odonatologists. C. *monomelaena* is northeastern in distribution, C. *fasciata* southern. If treated as one species, the range encompasses the entire eastern U. S. from Massachusetts west to Wisconsin, south to Texas and Florida. My Cape Cod records show its local range to extend to Chatham, but I do not doubt its existence on the Outer Cape.

Comments: As in Martha's Pennant, Banded Pennant breeding peaks in late July and early August. Adults often perch in the very tops of shoreline shrubs where they are well hidden from the collector. I walked the entire circumference of one Barnstable pond without spotting a single individual, but when I repeated the circuit with eyes focused on the shrubs I discovered dozens of Banded Pennants about six to eight feet up in highbush blueberry.

Pond Hawks (Genus *Erythemis*)

GREEN JACKET (*Erythemis simplicicollis*) **Plate 6**

Description: A medium-sized dragonfly, 39-44 mm. (1½-1¾ in.), whose appearance changes dramatically with age. Juvenile males and females boast bright lime green face and thorax. The abdomen, also a vivid lime green, is spotted brown on the middle segments. The last two segments are wholly brown and the male's superior appendages are yellow. The female's large triangular ovipositor points downward from beneath abdominal segment 9. Fully mature males become almost entirely frosty blue, with the exception of the green eyes and face and yellow superior appendages.

Flight Period: June through early September.

Range: New England, southern Quebec and Ontario west to Kansas, south to Texas and Florida. The Green Jacket is abundant from the Canal to the Cape-tip.

Comments: Green Jackets are savage predators, one of the few dragonflies which I regularly observe engaged in cannibalism. In one of the first incidents that sparked my interest in dragonflies, I watched a voracious adult Green Jacket methodically consume another of its own species from head to tail. Green Jackets perch on the ground, on logs or on low-lying plants. Males, while engaged in breeding competition, perform an intricate balletic circle dance: two individuals hover, one just in front of the other; the one behind flies up and over the leader who in turns dips down, back and up again so that the original positions of the two are reversed. The dragonflies continue this complex maneuver with remarkable agility and grace, the entire process appearing as one continuous fluid motion, like a wheel rolling along in mid-air. Green Jackets are common and widespread on Cape Cod, where they are found at a variety of habitats, from the larger sandy lakes to small mucky ponds and slow streams.

Dragonlets (Genus *Erythrodiplax*)

SALTMARSH DRAGONFLY (*Erythrodiplax berenice*) **Figure 27**

Description: I call this little dragonfly the Jekyll and Hyde of the saltmarsh, for its appearance changes dramatically with age. Teneral males and females are strikingly patterned with black and yellow zebra stripes on the thorax, and the top of the black abdomen is marked with bright orange spots. The face, prettily banded with black and yellow, is crowned with a shining metallic blue forehead. Males quickly darken, becoming completely blue-black in color, the bright markings on the abdomen obscured. Females too become black, but after a longer period of time. Wings of both sexes are clear, but there may be some amber color at the base or near the nodus. The female's scoop-shaped ovipositor projects downward from abdominal segment 9. Adults measure 30—33 mm. (1¼ in.).

Flight Period: Late June through August.

Range: Coastal: Quebec and Nova Scotia south to Florida, the Bahamas, Texas and Central America. It also occurs on the Pacific coast of Mexico and at a few inland salt lakes in the western states. The Saltmarsh Dragon is abundant Cape-wide, but restricted to coastal marshes.

Comments: This is the only dragonfly in North America found exclusively in saltwater habitats. On Cape Cod, the coastal salt and brackish marshes which it prefers are numerous, and the Saltmarsh Dragon is seldom found far from these habitats. Adults perch in marsh *Spartina* grasses, where they are well-hidden from view. On the high marsh, where salt hay forms low swirling windrows, they are more visible, flushing quickly to fly a few yards and light again. Saltmarsh Dragons feed on small insects, thus playing a significant role in mosquito control in marshes. Tandem pairs oviposit in stagnant ditches, where the eggs are enveloped in floating algae.

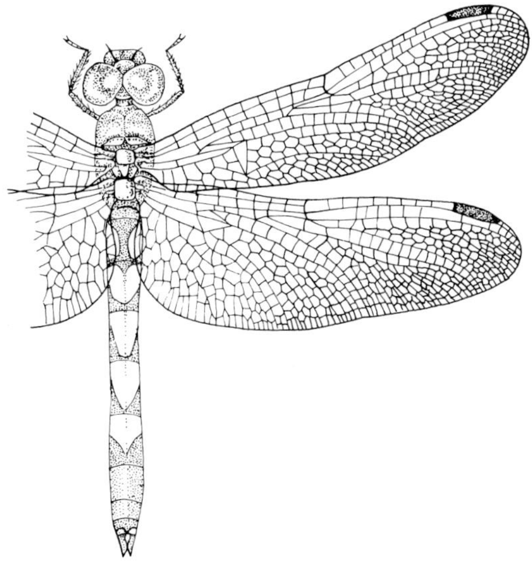

Figure 27. **Saltmarsh Dragonfly** (E. *berenice*), *teneral*.

Whitefaces (Genus *Leucorrhinia*)

The Whitefaces are mainly northern in distribution, there being only 2 representatives of the 7 North American species on Cape Cod. All are easily recognized by their creamy white faces, small size and dark bodies.

FROSTED WHITEFACE (*Leucorrhinia frigida*) **Not illustrated**

Description: A dainty little Whiteface at 28—30 mm. (less than 1¼ in.), L. *frigida* has no distinct color pattern. It is easily identified by its white face, black thorax and abdomen and, most importantly, the frosty grey pruinosity on the first four abdominal segments. There is a minute triangular black mark at the very base of the hind wings.

Flight Period: June and July.

Range: New England and Nova Scotia west to Manitoba, south only to North Dakota and Virginia. On Cape Cod, the Frosted Whiteface is recorded from Bourne and Falmouth to Hyannis.

Comments: This lovely little *Leucorrhinia* is much less abundant on Cape Cod than its relative, L. *intacta*. Like others of its genus, it prefers boggy ponds and is often associated with floating sphagnum bogs. It is thus rather limited in our area.

JOHNNY WHITEFACE (*Leucorrhinia intacta*) **Figure 28**

Description: A rather small dragonfly at 30-34 mm. (1¼-1⅜ in.), but one of the larger in its genus. Males are distinctive, with their pearly white face, hairy black thorax and black abdomen adorned with a bright yellow spot on top of segment 7. This spot, divided in two by a thin black line, is quite conspicuous against the black abdomen. There are also less noticeable spots on the sides of the first few segments which may be obscured in older individuals. The clear wings are marked only with tiny black basal spots. Females show considerably more pale abdominal spotting, with rectangular yellow marks on segments 3-6 and a small triangular spot on 7.

Flight Period: Late May to mid-July, most abundant in June.

Range: Generally northern: Nova Scotia and New England west through Canada to British Columbia south into California and east to Virginia. On Cape Cod, Bourne and Falmouth to Provincetown.

Comments: *Leucorrhinia intacta* inhabits shallow muddy or boggy ponds with plentiful floating and emergent vegetation. Males habitually perch on the floating leaves of waterlilies, to which they return regularly after making their rounds. The females are shy, retreating to surrounding woodlands, and are found along forest roads or in sunny openings in the canopy.

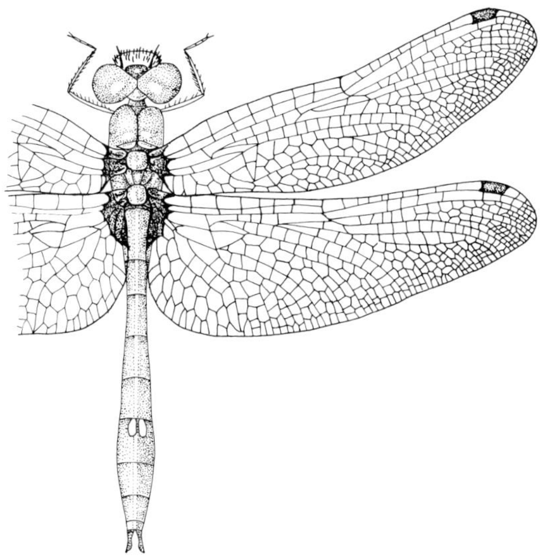

Figure 28. **Johnny Whiteface,** (L. *intacta*)

Common Skimmers (Genus *Libellula*)

These are our most common and indeed some of the most familiar summer dragonflies. They prefer still waters, and find the weedy shores of our shallow Coastal Plain Ponds ideal habitat. Many male Common Skimmers acquire a frosty blue or whitish pruinosity on the abdomen and sometimes the thorax at maturity. Twelve of the 22 North American species are found on Cape Cod.

GOLDENWINGS (*Libellula auripennis*) **Plate 7**

Description: The Goldenwings is aptly named for its dazzling orange color, which the male sports from head to tail and wingtip to wingtip. It is one of our largest Skimmers at 51-56 mm. (2-2¼ in.). The face, wing veins and abdomen of mature males are pumpkin colored, the stigmas crimson. Both eyes and thorax are rich chestnut brown. A narrow black dorsal stripe runs the entire length of the abdomen. Females are less conspicuous, clothed in protective brown like so many of this genus, with a black dorsal abdominal stripe as in the male.

Flight Period: Mid-June to early September.

Range: Eastern and southern: Massachusetts west to Missouri, south to Texas and Florida. On Cape Cod, Goldenwings are common from Falmouth to Provincetown.

Comments: This high-spirited, vigorously active summer species is abundant at grassy Coastal Plain Ponds. Males perch at the tips of emergent plants, usually Military Rush (*Juncus militaris*), from which they make broad forays about the territory. They are wary, very swift on the wing, and difficult to approach when at rest. Females, duller in color and more secretive in habit, oviposit alone while a male hovers nearby. Eggs are released by successive taps of the abdomen against the water.

BLACK-FACED SKIMMER (*Libellula cyanea*) **Figure 29**

Description: The Black-faced Skimmer is our only dragonfly with two-toned stigmas: half white and half black. Adults have a shining black face, brown thorax with creamy sides and pale gold abdomen with a broad black band on top from base to tip. At full maturity, males don a pale blue pruinosity on the thorax and abdomen. These are medium-sized dragonflies at 40-45 mm. (1½-1¾ in.). In addition to the unique bi-colored stigma, there is a faint wash of gold along the leading edge of the wing from the nodus outward, and a dark streak at the base of the wing.

Flight Period: June through August.

Range: Eastern: Maine south to Georgia, west to Texas, Kansas and Michigan. Widespread on Cape Cod, from Falmouth to Provincetown.

Comments: A pretty summer dragonfly which darts about the shores of shallow, muddy ponds, flashing its bold black and white stigmas and powdery blue abdomen. Males are territorial, returning to perch on a blade of grass or tall *Juncus* stem. Females oviposit alone in still shoreline water in the typical Skimmer fashion. The male hovers attentively nearby.

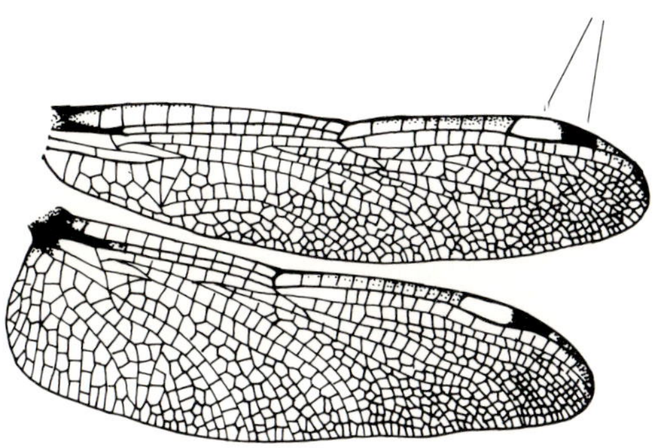

Figure 29. Wings of **Black-faced Skimmer** (L. *cyanea) showing two-toned stigmas.*

LITTLE CORPORAL (*Libellula deplanata*) **Not illustrated**

The following 2 Corporals are sometimes placed in a separate genus, called *Ladona*. In fact, the two are quite similar and considered one species by some Odonatologists. I believe they are distinctly different and thus separate species.

Description: The smaller of the 2 Corporals, at 32-36 mm. (1¼-1³/₈ in.). The Little Corporal bears two pale shoulder stripes on its dark brown thorax, hence the common name. Both the front of the thorax and the brown abdomen take on a uniform slatey blue pruinosity in mature males. The wings are clear with two dark brown streaks at the base. Females and immature males are brown, the abdomen with a black longitudinal stripe on top.

Flight Period: Spring, May and June.

Range: Southern: Massachusetts south to Florida west to Texas and Kansas. This species is widespread and common from Falmouth and Bourne to Provincetown.

Comments: The Little Corporal prefers quiet pond waters, but the habitat varies from the larger sandy kettles to small, grassy, mud-bottomed ponds. Adults habitually perch on the ground or on fallen logs. They may be seen sunning on woodland paths and in clearings during the early days of adult life.

CORPORAL SKIMMER (*Libellula exusta*) **Plate 7**

Description: This species is a larger, more robust version of L. *deplanata* at 34-40 mm. (1⅜-1⅝ in.), and easily distinguished from it by the more flashy white pruinosity of the mature male's abdomen. Unlike L. *deplanata*, the two pale corporal stripes of the thorax are not obscured by pruinosity. In most individuals the forewings show 2 short basal black streaks as in L. *deplanata*, while the hindwings have a roughly triangular mark at their base. Females and immature males are brown to reddish brown.

Flight Period: Spring, May to June.

Range: Northeastern: Nova Scotia and New Brunswick south through New England to Virginia. L. *exusta* is the more abundant of the 2 Corporals on Cape Cod. It is recorded from Bourne, Falmouth and Woods Hole to Provincetown.

Comments: A lively, energetic Skimmer, similar to L. *deplanata* in behavior, spring flight period and habitat requirements. In fact, the two coexist at many Cape Cod ponds, where they are often seen resting side by side on exposed pond shores.

DAMSON SKIMMER (*Libellula incesta*) **Plate 7**

Description: This is a large, dark Skimmer the color of a ripe plum, and is appropriately named for a variety of this sweet summer fruit. Adults measure 48—52 mm. (about 2 in.). Mature males are a deep, slatey midnight blue, the face and eyes nearly black. The wings, long and narrow, are clear with heavy black venation and stigmas. Females and teneral males are brownish, with creamy spots on the sides of the thorax and a continuous brown stripe running the length of the tawny abdomen. Segment 8 of the female's abdomen is expanded laterally, and there is usually some smoky brown color on the tips of the wings.

Flight Period: June through August.

Range: New England, southern Ontario, west to the Great Plains states, south to Texas and Florida. Abundant from Bourne and Sagamore to Provincetown.

Comments: The Damson is one of the most abundant and familiar of Cape Cod dragonflies. It is found at almost any Coastal Plain Pond, from the tiniest muddy quagmire to the largest sandy kettle, and even slow streams. Males are bold and territorial. They perch in shrubs along pond edges, dashing out to investigate intruders, competing males and potential mates. They do not hesitate to approach humans closely, and I often find myself "buzzed" by defensive individuals. Their flight involves quick darting maneuvers and frequent hovering. The dull brown females oviposit quietly in sheltered pond shore waters, hovering just above the surface and dipping the abdomen repeatedly into the water. The male "stands by," guarding. This species may be incredibly abundant at some ponds, with several hundred present.

WHITETAIL (*Libellula lydia*) **Plate 7**

Description: This is a chunky, medium-sized dragonfly, measuring 42-45 mm. (1¾ in.). Adults are strikingly handsome, virtually unmistakable insects. The face and thorax are deep brown, as is the abdomen of younger males and females. Two lateral pale thoracic stripes are visible in both sexes, but become obscured by thick hair in fully mature individuals. Lateral yellow spots on the male's abdomen are concealed by a brilliant white pruinosity at maturity. It is this trait for which the species is named. Each wing has a single broad black flag, beginning near the nodus and extending to the stigma. A smaller, less noticeable white spot adorns the base of the hind wings. The wings of females have 3 dark spots, one at the base, the midway point and the apex (Plate 7). They are very similar to female Tenspots (L. *pulchella*), from which they can be separated by the flattened expansion of abdominal segment 8.

Flight Period: Late May through early September.

Range: Throughout the U.S. into southern Canada and Mexico. On Cape Cod, the Whitetail is common and widespread, from Falmouth to Provincetown.

Comments: The familiar Whitetail inhabits freshwater ponds of various sizes, slow rivers and grassy marshes. After emergence, adults wander from the breeding habitat and are frequently seen resting on sunny roadways, paths or low-growing vegetation. Males, charismatic, spunky and fiercely territorial, intrepidly defend their piece of the pond and their mate. They charge swiftly about in command of the pond edge, driving off rivals and flashing their startling white abdomens as warning. Bold and spirited of flight, they are brazen enough to steal mates from other males. Mating requires only three seconds, a record for dragonflies. Males are well-known and indeed often studied for their zealous mate-guarding behavior. Though they seem fiercely loyal in guarding a mate, their attentions may wander to other females, at which times they will actually attempt to guard two mates at once! Though quite capable on the wing, the Whitetail's habit of resting on the ground makes it easy to capture. Several comprehensive studies of the mate-guarding behavior of this species have been done, and the riotous activity of breeding males continues to interest Odonatologists.

NEEDHAM'S SKIMMER (*Libellula needhami*) **Not illustrated**

Description: Needham's Skimmer is named for the famous Odonatologist James Needham. It is very similar to the Goldenwings in both size and color, but can be separated from this species in a number of ways: mature male Needham's Skimmers are more red in color, most of the wing veins are black and the legs are pale with black spines. An excellent diagnostic feature of both sexes is the costa vein on the leading edge of the wing: in Needham's Skimmer it is black from base to nodus, ochre from nodus to apex. In the Goldenwings, it is pale ochre from base to tip.

Flight Period: July and August.

Range: Coastal: Massachusetts south along the Atlantic to Florida and Texas, Mexico and the Caribbean. On Cape Cod, Needham's Skimmer is found in brackish tidal marshes, probably Cape-wide.

Comments: Despite its similarity to the Goldenwings in behavior, size and color, Needham's Skimmer is almost always found in brackish or saltwater marshes. On the Cape I have never seen it at fresh water, though it has been reported breeding in fresh habitats elsewhere.

TENSPOT (*Libellula pulchella*) **Plate 7**

Description: A large, conspicuous dragonfly, measuring 50-53 mm. (2 in.). Males, resplendent when fully mature, wear a rather uniform brown cloak when young. The head and thorax are warm chocolate brown, the thorax clothed with a fine pale pubescence. Two lateral yellow thoracic stripes are most evident in young individuals. The male's abdomen becomes grey blue with age, donning a powdery blue pruinosity not unlike that of the Whitetail, but of much less intensity. The ornate, harlequin wings of the Tenspot male are embellished with a remarkable black and white checkerboard pattern. Each wing has 3 black spots, one at the base, nodus and tip, with white alternating between. A total of 10 white spots on the 4 wings yields the descriptive name Tenspot, but the more obvious 12 black spots gives this species its second English name, Twelvespot. Females resemble the young male, but lack the bluish cast of the abdomen and white spots on the wings. See *Libellula lydia* for details on differences between females of these two species.

Flight Period: Early June through August.

Range: The Tenspot is a transcontinental species, occurring throughout the U.S. and southern Canada. On Cape Cod, it is recorded from Woods Hole, Falmouth and Bourne all the way to Provincetown.

Comments: This dashingly handsome Skimmer prefers shallow ponds where grasses and shoreline plants abound. It is particularly interesting both for its exotic colors and lively, animated behavior. Males are quite daring, at times approaching perilously close to the collector; but their speed, grace and finesse on the wing make them difficult to capture. Thus I am most often content just watching their antics on warm summer days. Mature males choose a conspicuous perch such as an overhanging branch or stick protruding several feet above the water, from which they have a wide view of their surroundings. Aggressive interactions, chases and scuffles with other dragonflies are numerous. I am often dive-bombed by aggressive males who fly directly from their perch to buzz my head, as though attempting to drive me from their territories! Females, quiet and elusive in contrast, drop eggs into pond shore waters, unattended by a mate.

PAINTED SKIMMER (*Libellula semifasciata*) — Figure 30

Description: A medium-sized dragonfly, measuring 40-47 mm. (1½-1¾ in.), with handsome ochre body color and distinctively spotted wings. The face becomes orange red at maturity, the eyes richly chestnut. Thorax and abdomen are golden brown with a hint of olive on the sides. The top of the last four abdominal segments and appendages are black. The wings are colored as follows: a wash of pale pumpkin and a black streak at the base, a darker brown spot at the nodus and a dipped-in-brown-ink effect at the wingtip. Females are similar to males.

Flight Period: June to early August.

Range: Eastern: Maine west through southern Ontario to Wisconsin, south through Kansas to Texas and Florida. L. *semifasciata* is widespread and common on Cape Cod. It is known from Falmouth and Woods Hole to Truro.

Comments: The Painted Skimmer is one of the first Libellulids on the wing, being most numerous at grassy Coastal Plain Ponds in June and early July. It is an ace on the wing and easily spooked, but perches often near the tops of emergent plants. It behaves much like the Goldenwings and in fact resembles that species on the wing. Females oviposit in the grassy shallows, tap-tap-tapping their abdomens on the water.

Elfin Skimmers (Genus N*annothemis*)

BLUEBELL (N*annothemis bella*) — Figure 31

Description: The tiniest of all North American Anisoptera, the Bluebell is a slender, delicate little creature, measuring only 20 mm. (3/4 in.). Mature males have a two-toned face, the lower half black, the upper creamy white. The coal black thorax acquires a greyish blue pruinosity with age, but the abdomen retains its blackness. The wings, clear with small black stigmas, have fewer veins and cells than other dragonflies. Females are black with yellow markings on the thorax and abdomen, and basal bands of yellow on the wings.

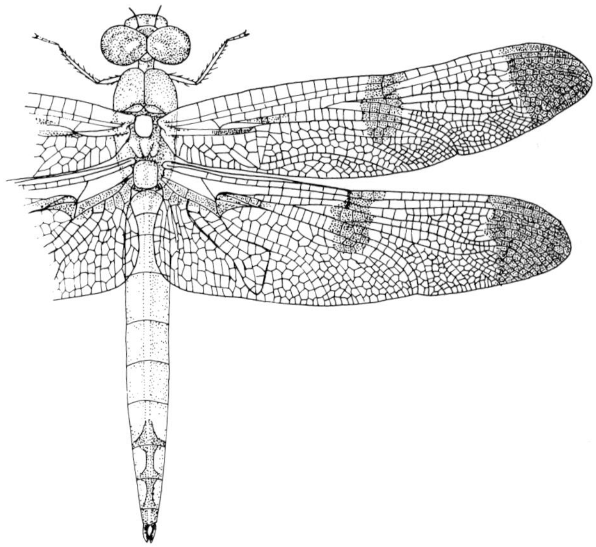

Figure 30. **Painted Skimmer** (L. *semifasciata*)

Flight Period: June and July.

Range: Eastern: New England, Quebec and Ontario west to Wisconsin, south to Louisiana and Florida. The Bluebell is known locally only from Provincetown, but has been reported historically from Falmouth and Brewster.

Comments: This extraordinarily tiny dragonfly is restricted to muddy quagmires and sphagnum bogs. It is quite rare on Cape Cod due to the scarcity of its sphagnum bog habitat. With its diminutive stature and relatively weak flight, the little Bluebell stays close to the sphagnum mat, seldom venturing beyond the bog edge. Because of its size, it should not be confused with any other species once it is spotted. Males are strongly territorial and surprisingly aggressive towards one another for so small a dragonfly.

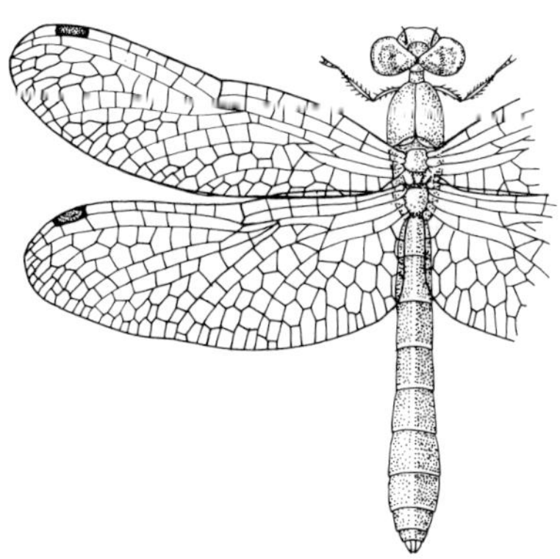

Figure 31. **Bluebell** (N. *bella*)

Blue Pirates (Genus *Pachydiplax*)

BLUE PIRATE or BLUE DASHER (*Pachydiplax longipennis*) **Plate 6**

Description: This medium-sized skimmer is unmistakable with its bright emerald eyes, creamy face, shining green forehead and slender powdery blue abdomen. Adults measure 36-40 mm. (1½ in.). The thorax sports an exotic zebra-striped pattern of dark brown and greenish yellow. The wing pattern is variable, but there is generally a splash of golden brown color at the base, overlaid on the hind wings with 3 short brown streaks. The outer half of the wings are washed with brown, very pale in some individuals or quite dark in others. The abdomen in mature males is a lovely pruinose blue like the bloom on ripe blueberries. Females are dark brown with pretty yellow stripes running lengthwise on the abdomen.

Flight Period: June through early September.

Range: Widespread, recorded from all lower 48 states (except North Dakota) to Mexico. The Blue Pirate is Cape-wide in distribution.

Comments: The Blue Pirate is a common, familiar dragonfly of quiet weedy ponds. Males are active, perky little fellows, dashing boldly about pond edges, tenaciously defending their territories. Hovering often, they are relatively easy to net. They perch frequently on twigs or emergent plants and can be seen on hot days with wings bent downward to shade the thorax. Females retire to surrounding woodlands until breeding. During oviposition, the female is guarded by her mate, who hovers overhead as she dunks her abdomen into the water.

Globetrotters or Rainpool Gliders (Genus *Pantala*)

There are two common species in the genus *Pantala*, and both occur on Cape Cod. Ecologically, they are unusual in that adults breed in temporary ponds. Nymphs grow rapidly in a race to mature before the aquatic habitat dries up. These are strong fliers with very broad wings and a distinctive gliding or sailing flight.

WANDERING GLOBETROTTER (*Pantala flavescens*) **Plate 8**

Description: A medium sized dragonfly with very long, broad wings and a large head relative to its body size. Adults are about 50 mm. (2 in.) long, and the wings are nearly as long as the body. The face is buffy yellow, becoming reddish in older individuals, the eyes rich chestnut. The pale greyish thorax bears broad, clear wings sometimes brushed with a faint flavescence at the base and near the tip. The tapered abdomen is deep golden yellow with brown spots atop its middle and end segments. The only similar species is the Spot-winged Globetrotter (*Pantala hymenea*), which is easily distinguished from *P. flavescens* by the single dark spot at the base of the hind wing (Figure 32).

Flight Period: June through early September.

Range: Cosmopolitan in distribution, *Pantala flavescens* breeds on all continents except Europe. In North America, it occurs from southeastern Canada west to Minnesota, Nevada and California, and throughout the southern U.S. *P. hymenea* ranges from southern Canada throughout the U.S., Central and South America. Locally, both species are common from the Canal to Provincetown.

Comments: Famous for its wanderlust, this Globetrotter drifts with the wind and weather, feeding on other wandering insects as it goes. Long, exceedingly broad wings are its mechanism for sustained, long-distance energy-efficient flight, and gliding or sailing is the preferred mode. The Wandering Globetrotter appears to have no fear of fickle ocean winds and is frequently spotted adrift on air currents hundreds of miles from land. The ultimate destiny of these ocean wanderers is a mystery, though it is likely that most never return to land. *Pantala flavescens* is a powerful, incredibly capable flier. They spend most of their time on the wing and despite the high frequency of gliding, they are very difficult to capture. I rarely see these dragonflies on the ponds, but more often find them feeding in fields, parking lots and along powerlines. Feeding in swarms, their flight is fast-paced with instant directional changes as they terrorize the green-heads and deer flies so annoying to humans.

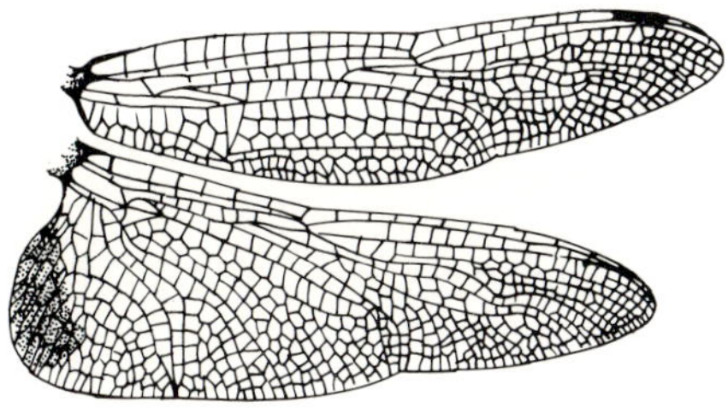

Figure 32. *Wings of* **Spot-winged Globetrotter** (P. *hymenea*)

Amberwings (Genus *Perithemis*

AMBERWINGS (*Perithemis tenera*) **Plate 8**

Description: A small, stout dragonfly with rather short abdomen and long wings. It is one of our smallest Anisoptera at 21-24 mm. (7/8-1 in.). Head, thorax and abdomen are dark brown with two pale greenish stripes showing through on the sides of the thorax. The wings, gracefully rounded on all edges, are the color of a pumpkin from base to tip. The stigmas are dark red, and some males have a faint or clearly visible dark brown spot near the triangles of each wing. The abdomen, short and rather oddly shaped, starts narrow, swells to its widest at segment 6 or 7, then tapers again to its tip. Females are similarly colored, but their wings are clear or diffusely ochre on the basal half, variously spotted with dark brown.

Flight Period: Late June through the end of August.

Range: Eastern: southern Ontario south to Florida, west to Texas and north to Nebraska. It is also known from northern Mexico. On Cape Cod, it ranges from Falmouth to Provincetown.

Comments: Amberwings fly low over the water's surface, avoiding the treacherous altitudes frequented by larger species. They perch often in low emergent vegetation. This species is frequently studied for its pre-mating courtship, a behavior which is easily observed at many Cape Cod ponds. Males first locate a suitable oviposition site which they defend from other males. A potential mate is guided to this spot and the male awaits her signal of acceptance, hovering nervously with his abdomen raised. If the female finds the site satisfactory, mating takes place and oviposition begins. With the male guarding, the female drops eggs onto floating debris or among emergent grasses, pond lilies and pickerelweed.

Meadowflies or Topers (Genus *Sympetrum*)

This is a large group of small, slender, reddish dragonflies, represented in North America by a dozen species, five of which have been recorded recently on Cape Cod. The

Meadowflies are typically found at grassy ponds or in nearby meadows, and some species fly very late into the autumn.

RUBY MEADOWFLY (*Sympetrum rubicundulum*) **Plate 8**

Description: One of the larger species of the genus, measuring 35-37 mm. (1⅜ in.). Males have a yellow face, reddish thorax with black legs, and bright crimson abdomen with black triangular markings on the sides. The red superior appendages bear a prominent tooth on the underside. Wings are clear, sometimes tinted with amber at the very base. Females are yellowish or brownish where males are red. In the very similar S. *internum*, the male's superior appendages are bright yellow.

Flight Period: Late June through October.

Range: Nova Scotia, southern Quebec and Ontario south to Georgia, west to Idaho and Utah. On Cape Cod, it is abundant in all towns.

Comments: This handsome Meadowfly is the most abundant and perhaps the most familiar species in its genus. On the day of emergence, hundreds of teneral adults rise in weak fluttery flight from pond shore grasses. Pairs oviposit in tandem, though on occasion I have seen females ovipositing alone: hovering about four feet in the air, they dropped clusters of eggs into stagnant pond waters. Post-breeding adults wander from the ponds and are often seen perched on garden flowers or sunning on warm walkways. They are not shy of humans and frequently alight on clothing, where they may stay for several minutes. This is one of the last dragonflies on the wing, often flying well into October or even November.

SAFFRON-BORDERED MEADOWFLY (*Sympetrum costiferum*) **Not illustrated**

Description: *Sympetrum costiferum* is the largest of our Meadowflies, at 36-38 mm. (1½ in.). It is uniformly red brown, with a golden yellow band covering the leading edge of the wing, from the base all the way to the yellow stigma. This band of color is darker in tenerals, but disappears almost completely in mature individuals. The thorax is greenish yellow on the sides, red brown above. The two-toned legs are yellow, striped with black on the outer sides.

Flight Period: July and August.

Range: Nova Scotia and the Maritimes to British Columbia south to California, east through Nebraska to New York and New England. Locally, S. *costiferum* is recorded from Barnstable, Mashpee, Brewster and Truro.

Comments: Despite its apparent restricted distribution on Cape Cod, I expect this species is more widespread than records indicate. Its habitat, shallow sandy ponds or grassy marshes, is certainly abundant here.

BAND-WINGED MEADOWFLY (*Sympetrum semicinctum*) **Not illustrated**

Description: An easily identified Meadowfly with bright red body color and orange-based wings. Adults are rather small at 28-31 mm. (1⅛-1¼ in.), with reddish brown face and thorax and cherry red abdomen. The superior appendages of the male are red, the inferiors black. The most distinctive feature of this species is its wings, the basal halves of which are banded with bright orange. This band of color, heaviest on the hind wings, is paler and reduced in area on the forewings.

Flight Period: July through September.

Range: Nova Scotia, New Brunswick west to Wisconsin, south and east to North Carolina. On Cape Cod, I have only one current location, in Wellfleet, although historical records also exist for Mashpee and Barnstable.

Comments: This relatively scarce Meadowfly is usually seen in the vicinity of running water. In Wellfleet, it was quite abundant along the mucky shores of Herring Pond, within a few yards of the Herring River. Single males and tandem pairs perch low in tall grasses and oviposit in stagnant pools along the shoreline.

YELLOW-LEGGED MEADOWFLY (*Sympetrum vicinum*) **Plate 8**

Description: A dainty little *Sympetrum* (32 mm. or 1¼ in.) with orange red face, reddish brown thorax and orange red abdomen. There are no black marks on the abdomen as in S. *rubicundulum*. The legs, red at the base, become pale yellow beyond. The clear wings may be brushed with a bit of amber color at the base, and the stigmas are red. Females are similar to males.

Flight Period: Late July through early October.

Range: Nova Scotia south to Georgia, west to Colorado, north to Wisconsin and Ontario. Also from Washington and British Columbia. *Sympetrum vicinum* is common from Bourne to Provincetown.

Comments: Like *Sympetrum rubicundulum*, this autumnal species becomes incredibly abundant in places. It emerges somewhat later than its relative and flies well into October.

Saddlebags or Dancing Gliders (Genus *Tramea*)

Two of the 7 North American *Trameas* occur on Cape Cod. They are similar in shape to Globetrotters, but have a dark band at the base of the hind wings, resembling the saddlebags of western riding gear. A curious egg-laying behavior has earned members of this genus the English name Dancing Gliders: as the tandem pair skims the pond surface, the female releases eggs by dipping her abdomen into the water. If you watch very closely, you can see the male release the female at the moment of oviposition, then grasp her again, repeating this interesting maneuver as egg-laying continues.

RED SADDLEBAGS (*Tramea carolina*) **Plate 8**

Description: A resplendent red dragonfly with very long broad wings much like the Globetrotters. It is one of the prettiest of our Skimmers, and at 48-52 mm. (2 in.), also one of the largest. The face, rufous in mature individuals, is crowned with a metallic violet forehead. The thorax is reddish brown and downy, the abdomen brilliant ruby red, with the exception of segments 8 and 9, which are black. The hind wings, broad-based and triangular, bear a characteristic dark basal spot which covers the entire width of the wing, extending nearly ⅓ its length. Both wing venation and stigmas are reddish brown. The Black Saddlebags (*Tramea lacerata*) which also occurs in our area, is very similar, but mature individuals are black from head to tip of abdomen, including the wide basal hind wing marks.

Flight Period: June through August

Range: Eastern: New England, southern Ontario west to Wisconsin and Kansas, south to Texas and Florida. I have seen *Tramea carolina* all over the Cape, but it is most common between the Canal and Orleans.

Comments: *Tramea carolina* is found at shallow, grassy Coastal Plain Ponds with thick emergent and underwater vegetation. It flies with a buoyant, bouncing motion, gliding methodically along the shoreline 2 or 3 yards above the water. Males choose an exposed perch from which to survey their kingdom. Mated pairs are often seen in tandem and ovipositing on the ponds. The first Saddlebags I caught so impressed me with its rich red body and shining violet face, and the feel of pulsing flight muscles struggling to be free that, in a weak moment, I gently tossed him into the air and watched him bounce away.

Species Checklist of Cape Cod Dragonflies and Damselflies

The following list reflects all those species recorded on Cape Cod during the years of my study, and those recorded by other Odonatologists who collected locally. Historic records generally date from the early part of this century to the 1950s. My thanks to Paul Miliotis for information on the very rare Brotherly Clubtail he collected on Cape Cod. It is likely that additional species will be discovered as the years progress.

Damselflies (Suborder Zygoptera)

Broad-winged Damselflies (Family Calopterygidae)

Ebony Jewelwing *Calopteryx maculata**
American Rubyspot *Hetaerina americana*

Spread-winged Damselflies (Family Lestidae)

Spotted Spreadwing *Lestes congener*
Disjunct Spreadwing *L. disjunctus*
Emerald Spreadwing *L. dryas* (historic)
Amber-winged Spreadwing *L. eurinus**
Sweetflag Spreadwing *L. forcipatus*
Elegant Spreadwing *L. inaequalis*
Slender Spreadwing *L. rectangularis**
Lyre-tipped Spreadwing *L. unguiculatus* (historic)
Swamp Spreadwing *L. vigilax**

Narrow-winged Damselflies (Family Coenagrionidae)

Eastern Red Damsel *Amphiagrion saucium**
Powdered Dancer *Argia moesta**
Violet Dancer *A. fumipennis**
Variegated Damselfly *Chromagrion conditum**
Bog Bluet *Enallagma aspersum**
Civil Bluet *E. civile**
Common or Northern Bluet *E. cyathigerum**
Turquoise Bluet *E. divagans*
Doubleday's Bluet *E. doubledayi**
Big Bluet *E. durum* (historic)
Stream Bluet *E. exsulans*
Skimming Bluet *E. geminatum**
Hagen's Bluet *E. hageni* (historic)
Lateral Bluet *E. laterale**
Little Bluet *E. minusculum**
Red Bluet *E. pictum**
Barrens Bluet *E. recurvatum**
Orange Bluet *E. signatum**

Slender Bluet E. *traviatum*
Evening Bluet E. *vesperum**
Speartail *Ischnura hastata**
Kellicott's Forktail I. *kellicotti*
Fragile Forktail I. *posita**
Rambur's Forktail I. *ramburii* (historic)
Common Forktail I. *verticalis**
Sphagnum Sprite *Nehalennia gracilis**
Sedge Sprite N. *irene**

Dragonflies (Suborder Anisoptera)

Darners (Family Aeshnidae)

Canada Blue Darner *Aeshna canadensis*
Spotted Blue Darner A. *clepsydra**
Spring Blue Darner A. *mutata*
Black-tipped Darner A. *tuberculifera**
Shadow Darner A. *umbrosa**
Eastern Blue Darner A. *verticalis*
Common Green Darner *Anax junius**
Long-legged Green Darner A. *longipes**
Hero Darner or Swamp Darner *Epiaeschna heros**
Springtime Darner *Basiaeschna janata**
Stream Darner *Boyeria vinosa**
Harlequin Bog Darner *Gomphaeschna furcillata**

Clubtails (Family Gomphidae)

Black-shouldered Spinyleg *Dromogomphus spinosus**
Abbreviated Clubtail *Gomphus abbreviatus* (historic)
Lancet Clubtail G. *exilis**
Brotherly Clubtail G. *fraternus*
Dusky Clubtail G. *spicatus* (historic)
Sand Dragon *Progomphus obscurus**

Emeralds (Family Corduliidae)

Petite Emerald *Dorocordulia lepida**
Common Baskettail *Epitheca (Tetregoneuria) cynosura**
Prince or Royal Skimmer E. *princeps**
Clamp-tipped Emerald *Somatochlora tenebrosa**
Broomtail S. *walshii*

Belted and River Skimmers, or Cruisers (Family Macromiidae)

Belted Skimmer or Stream Cruiser *Didymops transversa**
River Skimmer or Swift River Cruiser *Macromia illinoiensis**

Common Skimmers (Family Libellulidae)

Elisa Skimmmer *Celithemis elisa**
Halloween Pennant C. *eponina**

Martha's Pennant C. *martha**
Banded Pennant C. *monomelaena**
Green Jacket *Erythemis simplicicollis**
Saltmarsh Dragonfly *Erythrodiplax berenice**
Frosted Whiteface *Leucorrhinia frigida**
Johnny Whiteface L. *intacta**
Goldenwings *Libellula auripennis**
Black-faced Skimmer L. *cyanea**
Little Corporal L. *deplanata**
Corporal Skimmer L. *exusta**
Damson Skimmer L. *incesta**
The Widow L. *luctuosa*
Whitetail L. *lydia**
Needham's Skimmer L. *needhami**
Tenspot L. *pulchella**
Four-Spotted Skimmer L. *quadrimaculata* (historic)
Painted Skimmer L. *semifasciata**
Great Blue Skimmer L. *vibrans*
Bluebell *Nannothemis bella**
Blue Pirate or Blue Dasher *Pachydiplax longipennis**
Wandering Globetrotter *Pantala flavescens**
Spot-winged Globetrotter P. *hymenea**
Amberwings *Perithemis tenera**
Variegated Meadowfly *Sympetrum corruptum* (historic)
Ruby Meadowfly S. *rubicundulum**
Saffron-bordered Meadowfly S. *costiferum**
Band-winged Meadowfly S. *semicinctum**
Cherry-faced Meadowfly S. *internum**
White-faced Meadowfly S. *obtrusum* (historic)
Yellow-legged Meadowfly S. *vicinum**
Red Saddlebags *Tramea carolina**
Black Saddlebags T. *lacerata**

*Described in text—see Index of Scientific Names and Index of Common Names

Suggested Readings

Corbet, Philip S. 1963. A *Biology of Dragonflies*. Quadrangle Books, Inc., Chicago. 247 pp.

Dunkle, Sidney W. 1989. *Dragonflies of the Florida Peninsula, Bermuda and the Bahamas*. Scientific Publishers Nature Guides. 154 pp.

— 1990. *Damselflies of Florida, Bermuda and the Bahamas*. Scientific Publishers Nature Guides. 148 pp.

Garman, Philip. 1927. *The Odonata or Dragonflies of Connecticut*. State Geological and Natural History Survey, Hartford. 331 pp. Out of print.

McCafferty, W. Patrick. 1981. *Aquatic Entomology*. Science Books International, Inc. 448 pp.

Needham, James G. and Hortense B. Heywood. 1929. A *Handbook of Dragonflies of North America*. Charles Thomas, Springfield, Illinois. 378 pp. Out of print.

Needham, James G. and Minter S. Westfall, Jr. 1954. *The Dragonflies of North America*. University of California Press. 615 pp.

Walker, Edmund M. and Philip S. Corbet. 1975. *The Odonata of Canada and Alaska, Vol. III: The Anisoptera—Three Families*. University of Toronto Press. 308 pp.

Index to Scientific Names

This index lists the scientific names of all species and other taxonomic classifications described in the text. **Boldface** page or plate numbers indicate illustrations.

Index to Common Names

Boldface page and plate numbers indicate illustrations. Alternate common names are also given, followed by parentheses which direct the reader to the name under which the species is treated in the text.